Mechanical Engineering Dictionary

Mechanical Engineering Dictionary

Dr. Ramesh Chandra Nayak
Dr. Manmatha Kumar Roul

CWP

Central West Publishing

This edition has been published by Central West Publishing, Australia
© 2021 Central West Publishing

For more information about the books published by Central West Publishing, please visit https://centralwestpublishing.com

Disclaimer
Every effort has been made by the publisher and author while preparing this book, however, no warranties are made regarding the accuracy and completeness of the content. The publisher and author disclaim without any limitation all warranties as well as any implied warranties about sales, along with fitness of the content for a particular purpose. Citation of any website and other information sources does not mean any endorsement from the publisher and authors. For ascertaining the suitability of the contents contained herein for a particular lab or commercial use, consultation with the subject expert is needed. In addition, while using the information and methods contained herein, the practitioners and researchers need to be mindful for their own safety, along with the safety of others, including the professional parties and premises for whom they have professional responsibility. To the fullest extent of law, the publisher and author are not liable in all circumstances (special, incidental, and consequential) for any injury and/or damage to persons and property, along with any potential loss of profit and other commercial damages due to the use of any methods, products, guidelines, procedures contained in the material herein.

NATIONAL LIBRARY OF AUSTRALIA

A catalogue record for this book is available from the National Library of Australia

ISBN (print): 978-1-922617-23-1

About the Authors

Dr. Ramesh Chandra Nayak is presently working as the Professor and Head of Mechanical Engineering Department at Synergy Institute of Technology, Bhubaneswar, Odisha, India. He has 15 years of teaching, research and administrative experience. He has accomplished his B. Tech. and M. Tech. in Mechanical Engineering from BPUT, Odisha, India, with Ph.D. in Mechanical Engineering with a specialization in Thermal Engineering from SOA University, Bhubaneswar, Odisha, India.

Dr. Manmatha Kumar Roul is presently working as the Professor and Principal at GITA Autonomous College, Bhubaneswar, India. He has 30 years of teaching, research and administrative experience. He has attained his B.Tech. in Mechanical Engineering from VSSUT Burla, India, with M.Tech. and Ph.D. in Mechanical Engineering with a specialization in Thermal Engineering from IIT Kharagpur, India.

Preface

Mechanical engineering is one of the oldest and broadest branches of engineering concerned primarily with applications of the principles of physics and mathematics for designing, analysing, production and operation of various machines and systems. The field of mechanical engineering requires an understanding of core areas including mechanics, thermodynamics, materials science, heat transfer, refrigeration and air conditioning, aircraft propulsion, robotics, machine dynamics, fluid mechanics, structural analysis, etc. This dictionary provides definitions and explanations for various mechanical engineering terms in the core areas of thermodynamics, fluid mechanics, mechanics, manufacturing, heat transfer, design and other areas related to mechanical engineering in over 4500 clear and brief A to Z entries. This book will be helpful for students to create their early beginnings towards their career in core areas as well as higher studies.

We would like to thank everyone who has cooperated with us and encouraged us in producing this book in a very short span of time.

A **Able**: Having sufficient power, skill, or resources to do something.

Abrasion test: The test for hardness property of aggregates.

Abrasive belt: A cloth, leather, or paper band impregnated with grit and rotated as an endless loop to abrade materials through unremitting friction.

Abrasive blasting: The cleaning or finishing of surfaces by the use of an abrasive entrained in a blast of air.

Abrasive disc: The abrasive discs are a fundamental piece to be able to use a fixed or portable tool that helps us to cut, roughing or sanding any material, through friction.

Abrasive jet cleaning: The removal of dirt from a solid by a gas or liquid jet carrying abrasives to ablate the surface.

Abrasive machining: Process where material is removed from a work piece using a multitude of small abrasive particles. Example: Grinding, drilling, shaping, or polishing by abrasion.

Abreast milling: A milling method in which parts are placed in a row parallel to the axis of the cutting tool and are milled simultaneously.

Absolute specific gravity: The ratio of the weight of a given volume of a substance in a vacuum at a given temperature to the weight of an equal volume of water in a vacuum at a given temperature.

Absorption cycle: In refrigeration, the process whereby a circulating refrigerant, for example, ammonia, is evaporated by heat from an aqueous solution at elevated pressure and subsequently reabsorbed at low pressure, displacing the need for a compressor.

Absorption refrigeration: Refrigeration in which cooling is effected by the expansion of liquid ammonia into gas and absorption of the gas by water; the ammonia is reused after the water evaporates.

Absorption system: The refrigeration system where compressor is replaced by an absorber, in this system the refrigerant gas in the evaporator is taken up by an absorber and is then, with the application of heat, released in a generator.

Absolute pressure: It is the zero reference to a perfect vacuum, which exists in the air free space of the universe. It is denoted by Abs. Absolute pressure is the sum of gauge pressure and atmospheric pressure.

Acceleration of gravity: The acceleration imparted to bodies by the attractive force of the earth.

Accelerator: The device for varying the speed of an automotive vehicle with varying the supply of fuel.

Accelerator jet: The jet through which the fuel is injected into the incoming air in the carburetor of an automotive vehicle with rapid demand for increased power output.

Accelerator linkage: The linkage connecting the accelerator pedal of an automotive vehicle to the carburetor throttle valve or fuel injection control.

Accelerator pedal: A pedal that operates the carburetor throttle valve or fuel injection control of an automotive vehicle.

Accelerator pump: A small cylinder and piston controlled by the throttle of an automotive vehicle so as to provide an enriched air- fuel mixture during acceleration.

Accessory: A part, subassembly, or assembly, that are not usually essential, but which can be used to make the work more efficient, accessories are used for testing, adjusting, calibrating, recording, or other purposes.

Actual power: The power calculated at the output shaft of a system.

Actuated roller switch: A centrifugal sequence-control switch which is placed in contact with a belt conveyor.

Adamantine drill: The Adamantite Drill is a Hardmode drill. It is employed in rotary drilling in very hard ground.

Adhesion: Sticking metal surfaces together under compressive stresses by formation of metallic bonds.

Adhesive bond: The forces such as dipole bonds which attract adhesives and base materials to each other.

Adiabatic engine: A heat engine in which there is no gain or loss of heat.

Adiabatic envelope: A surface enclosing a thermodynamic system, when no heat flows through the surface.

Adiabatic expansion: Increase in volume without heat flow, in or out.

Adiabatic process: A process which takes place in a system without the exchange of heat with the surroundings.

Adiabatic vaporization: Vaporization of a liquid with no heat exchange between it and its surroundings.

Adsorption system: A device that dehumidifies air by bringing it into contact with a solid adsorbing matter.

Aerial spud: A cable for moving and anchoring a dredge.

Aerial tramway: A system for transporting bulk materials that consists of one or more cables supported by steel towers and is capable of carrying a travelling carriage from which loaded buckets can be lowered or raised.

Aeroballistics: The study of the interaction of projectiles or high-speed vehicles with the atmosphere.

Aerodynamic trajectory: A trajectory or element of a trajectory in which the missile or vehicle encounters enough air resistance to stabilize its flight or to modify its course significantly.

Aero elasticity: The deformation of structurally elastic bodies in response to aerodynamic loads.

Aero fall mill: A grinding mill of large diameter with either lumps of ore, pebbles, or steel balls as crushing bodies.

After running: In an automotive engine, continued operation of the engine after the ignition switch is turned off.

After top dead center: The position of the piston after reaching the top of its stroke in an automotive engine.

Agitating speed: The rate of rotation of the drum or blades of a truck mixer or other device used for agitation of mixed concrete.

Agitator: A device for keeping liquids and solids in liquids in motion by mixing, stirring, or shaking.

Air: Medium, a mixture mainly of oxygen and nitrogen.

Air aspirator valve: On certain automotive engines, a one-way valve installed on the exhaust manifold to allow air to enter the exhaust system; provides extra oxygen to convert carbon monoxide to carbon dioxide. Also known as gulp valve.

Air bag: An automotive vehicle passenger safety tool consisting of a passive restraint in the form of a bag which is automatically inflated with gas to provide cushioned protection against the impact of a collision.

Air bleeder: A device, such as a needle valve, for removing air from a hydraulic system.

Air brake: An energy conversion mechanism activated by air pressure and used to retard, stop, or hold a vehicle or, generally, any moving element.

Air cap: A tool used in thermal spraying which directs the air pattern for purposes of atomization.

Air cell: A small auxiliary combustion chamber used to help turbulence and develop combustion in certain types of diesel engines.

Air chamber: A pressure vessel, partially filled with air, for converting pulsating flow to steady flow of water in a pipeline, as with a reciprocating pump.

Air classifier: A device to break up particles by size through the action of a stream of air.

Air compressor: A machine that increases the pressure of air by increasing its density.

Air-compressor unloader: A device for control of air volume flowing through an air compressor.

Air-compressor valve: A tool for controlling the flow into or out of the cylinder of a compressor.

Air conditioner: A mechanism primarily for comfort cooling that lowers the temperature and reduces the humidity of air in buildings.

Air condenser: A steam condenser in which the heat exchange occurs through metal walls separating the steam from cooling air.

Air conditioner: A mechanism primarily for comfort cooling that lowers the temperature and reduces the humidity of air in buildings.

Air conditioning: The maintenance of certain aspects of the environment within a defined space to facilitate the function of that space; aspects controlled include air temperature and motion, radiant heat level, moisture, and concentration of pollutants such as dust, microorganisms, and gases. Also known as climate control.

Air-cooled engine: An engine cooled directly by a stream of air without the interposition of a liquid medium.

Air-cooled heat exchanger: A finned-tube (extended-surface) heat exchanger with hot fluids inside the tubes, and cooling air that is fan-blown (forced draft) or fan-pulled (induced draft) across the tube bank.

Air cooling: Lowering of air temperature for comfort, process control, or food preservation.

Air cycle: A refrigeration cycle characterized by the working fluid, air, remaining as a gas throughout the cycle rather than being condensed to a liquid; used primarily in airplane air conditioning.

Air cylinder: A cylinder in which air is compressed by a piston, compressed air is stored, or air drives a piston.

Air density: The mass per unit volume of air.

Air engine: An engine in which compressed air is the actuating fluid.

Air-fuel mixture: In a Spark ignition engine, the charge of air and fuel that is mixed in the appropriate ratio in the carburetor.

Air-handling system: An air-conditioning system in which an air handling unit provides part of the treatment of the air.

Air horsepower: The theoretical (minimum) power required to deliver the specified quantity of air under the specified pressure conditions in a fan, blower, compressor, or vacuum pump.

Air-injection system: A device that uses compressed air to inject the fuel into the cylinder of an internal combustion engine.

Air lift: Tools for lifting slurry or dry powder through pipes by means of compressed air.

Air line: A fault, in the form of an elongated bubble, in glass tubing.

Air motor: A device in which the pressure of cramped air causes the rotation of a rotor or the movement of a piston.

Air nozzle: Air nozzle is a device used control the direction or characteristics of air flow by converting pressure into flow.

Air preheater: A device used in steam boilers to transfer heat from the flue gases to the combustion air before the matter enters the furnace.

Air pump: A device for removing air from an enclosed space or for adding air to an enclosed space.

Air receiver: A vessel designed for compressed-air installations that is used both to store the compressed air and to permit pressure to be equalized in the system.

Air reheater: In a heating system, any device used to add heat to the air circulating in the system.

Air resistance: Wind drag giving rise to forces and wear on buildings and other structures.

Air-standard cycle: A thermodynamic cycle in which the working fluid is considered to be a perfect gas

Air-standard engine: A heat engine operated in an air-standard cycle.

Air system: A mechanical refrigeration system in which air serves as the refrigerant in a cycle comprising compressor, heat exchanger, expander, and refrigerating core.

Air valve: A valve that automatically lets air out of or into a liquid-carrying pipe when

the internal pressure drops below atmospheric.

Analytic mechanics: The application of differential and integral calculus to classical (nonquantum) mechanics.

Analyzer: A multifunction test meter, measuring volts, ohms, and amperes. Also known as set analyzer.

Andrade's creep law: A law which states that creep exhibits a transient state in which strain is proportional to the cube root of time and then a steady state in which strain is proportional to time.

Andrews's curves: A series of isotherms for carbon dioxide, showing the dependence of pressure on volume at various temperatures.

An elasticity: An elasticity is a unit free measure. Deviation from a proportional relationship between stress and strain.

Aneroid valve: A valve actuated or controlled by an aneroid capsule.

Angle of action: The angle of revolution of either of two wheels in gear during which any particular tooth remains in contact.

Angle back-pressure valve: A back pressure valve with its outlet opening at right angles to its inlet opening.

Angle of impact: The acute angle between the tangent to the trajectory at the point of impact of a projectile and the plane tangent to the surface of the ground or target at the point of impact.

Angle of nip: The largest angle that will just grip a lump between the jaws, rolls, or mantle and ring of a crusher. Also known as angle of bite; nip.

Angle of orientation: Of a projectile in flight, the angle between the plane determined by the axis of the projectile and the tangent to the trajectory (direction of motion), and the vertical plane including the tangent to the trajectory.

Angle of pressure: The angle between the profile of a gear tooth and a radial line at its pitch point. Also known as angle of obliquity.

Angle of recess: The angle that is turned through by either of two wheels in gear, from the coincidence of the pitch points of a pair of teeth until the last point of contact of the teeth.

Angle of repose or Angle of rest: The angle between the horizontal and the plane of contact between two bodies when the upper body is just about to slide over the lower. Also known as angle of friction.

Angle of torsion: The angle through which a part of an object such as a shaft or wire is rotated from its normal position when a torque is applied.

Angular milling: Milling surfaces that are flat and at an angle to the axis of the spindle of the milling machine.

Angular momentum: The product of mass, velocity and radius.

Angular speed: Angular speed is the speed of the object in rotational motion. The speed which is the linear speed = angular speed X radius of the rotation.

Antifriction: Making friction smaller in magnitude.

Antifriction bearing: Any bearing having the capability of reducing friction effectively.

Antilock braking system: For vehicles, a sensor-control system found in braking systems which prevents wheel lockup while allowing the brakes to continue slowing the wheel. Abbreviated ABS.

Approach: The difference between the temperature of the water leaving a cooling tower and the wet-bulb temperature of the surrounding air.

Apparent force: A force introduced in a relative coordinate system in order that Newton's laws be satisfied in the system; examples are the Coriolis force and the centrifugal force incorporated in gravity.

Apparent weight: The difference between the temperature of the water leaving a cooling tower and the wet-bulb temperature of the surrounding air.

Approach: For a body immersed in a fluid (such as air), the resultant of the gravitational force and the buoyant force of the fluid acting on the body; equal in magnitude to the true weight minus the weight of the displaced fluid.

Apron conveyor: A conveyor used for carrying granular or lumpy material and consisting of two strands of roller chain separated by overlapping plates.

Apron feeder: A limited-length version of apron conveyor used for controlled-rate feeding of pulverized materials to a process or packaging unit.

Arbour: A cylindrical device positioned between the spindle and outer bearing of a milling machine and designed to hold a milling cutter.

Arbor hole: A hole in a revolving cutter or grinding wheel for mounting it on an arbor.

Arbor press: A machine used for forcing an arbor or a mandrel into drilled or bored parts preparatory to turning or grinding.

Archimedes' screw: A device for raising water by means of a rotating broad-threaded screw or spirally bent tube within an inclined hollow cylinder.

Ash conveyor: Ash conveyors are devices that help move ash from one system to the next. The most common of these conveyors use a giant screw inside a tube.

Aspect ratio: In any rectangular configuration, the ratio of the longer dimension to the shorter.

Assembly: A unit containing the component parts of a mechanism, machine, or similar device.

Assembly machine: A machine in a manufacturing facility that produces a configuration of some practical value from discrete components.

Asymmetric rotor: A rotating part for which the axis (center of rotation) is not centered in the element.

Asymmetric top: A system in which all three principal moments of inertia are different.

Ata: A unit of absolute pressure in the metric technical system equal to 1 technical atmosphere.

Atmospheric pressure: The pressure exerted by the atmosphere. A unit of pressure equal to101.325 kilopascals, which is the air pressure measured at mean sea level.

Atmospheric cooler: A fluids cooler that utilizes the cooling effect of ambient air surrounding the hot, fluids-filled tubes.

Atomization: Atomization is the process of converting an analyte in solid, liquid or solution form to a free gaseous atom. It is the transformation of a bulk liquid into a spray of liquid droplets in a surrounding gas or vacuum.

Atomizer: A device that produces a mechanical subdivision of a bulk liquid, as by spraying, sprinkling, misting, or nebulizing.

Atomizer burner: A liquid-fuel burner that atomizes the un ignited fuel into a fine spray as it enters the combustion zone.

atomizing humidifier: A humidifier in which tiny particles of water are introduced into a stream of air.

Auger: A wood-boring tool that consists of a shank with spiral channels ending in two spurs, a central tapered feed screw, and a pair of cutting lips.

Auger packer: A feed mechanism that uses a continuous auger or screw inside a cylindrical sleeve to feed hard-to-flow granulated solids into shipping containers, such as bags or drums.

Autoignition: Spontaneous ignition of some or all of the fuel-air mixture in the combustion chamber of an internal combustion engine. Also known as spontaneous combustion.

Automatic batcher: A batcher for concrete which is actuated by a single starter switch, opens automatically at the start of the weighing operations of each material, and closes automatically when the designated weight of each material has been reached.

Automatic drill: A straight brace for bits whose shank comprises a coarse-pitch screw sliding in a threaded tube with a handle at the end; the device is operated by pushing the handle.

Automatic fire pump: A pump which provides the required water pressure in a fire standpipe or sprinkler system; when the water pressure in the system drops below a preselected value, a sensor causes the pump to start.

Automatic press: A press in which mechanical feeding of the work is synchronized with the press action.

Automatic sampler: A mechanical device to sample process streams (gas, liquid, or solid) either continuously or at preset time intervals.

Automatic screw machine: A machine designed to automatically produce finished parts from bar stock at high production rates; the term is not an exact, specific machine tool classification.

Automatic stoker: A device that supplies fuel to a boiler furnace by mechanical means. Also known as mechanical stoker.

Automobile: A passenger vehicle designed for operation on ordinary roads and typically having four wheels and a gasoline or diesel internal combustion engine.

Automobile chassis: The automobile frame, together with the wheels, power train, brakes, engine, and steering system.

Automotive air conditioning: A system for maintaining comfort of occupants of automobiles, buses, and trucks, limited to air cooling, air heating, ventilation, and occasionally dehumidification.

Automotive brake: A friction mechanism that slows or stops the rotation of the wheels of an automotive vehicle, so that tire traction slows or stops the vehicle.

Automotive ignition system: A device in an automotive vehicle which commence the chemical reaction between fuel and air in the cylinder charge.

Automotive steering: by which a driver controls the course of a moving automobile, bus, truck, or tractor.

Automotive suspension: The springs and related parts intermediate between the wheels and frame of an automotive vehicle that support the frame on the wheels and absorb road shock caused by passage of the wheels over irregularities.

Automotive transmission: A device for providing different gear or drive ratios between the engine and drive wheels of an automotive vehicle, a principal function being to enable the vehicle to accelerate from rest through a wide speed range while the engine operates within its most effective range.

Automotive vehicle: A self-propelled vehicle or machine for land transportation of people or commodities or for moving materials, such as a passenger car, bus, truck, motorcycle, tractor, airplane, motorboat, or earthmover.

Automotive engine: The fuel consuming machine that provides the motive power for automobiles, airplanes, tractors, buses, and motorcycles and is carried in the vehicle.

Automotive engineering: The branch of mechanical engineering concerned primarily with the special problems of land transportation by a four-wheeled, trackless, automotive vehicle.

Autorail: A self-propelled vehicle having both flange wheels and pneumatic tires to allow operation on both rails and roadways.

Autorotation: Rotation about any axis of a body that is symmetrical and exposed to a uniform airstream and maintained only by aerodynamic moments. Rotation of a stalled symmetrical airfoil parallel to the direction of the wind.

Autosled: it is a technology used in transport management system. A propeller-driven machine equipped with runners and wheels and adaptable to use on snow, ice, or bare roads.

Auxiliary power plant: Ancillary equipment, such as pumps, fans, and soot blowers, used with the main boiler, turbine, engine, or generator of a power-generating station.

Availability: The probability that a system is operating satisfactorily at any point in time, excluding times when the system is under repair.

Availability ratio: The ratio of the amount of time a system is actually available for use to the amount of time it is supposed to be available.

Available draft: The usable differential pressure in the combustion air in a furnace, used to sustain combustion of fuel or to transport products of combustion.

Available energy: Energy which can in principle be converted to mechanical work.

Available heat: The heat per unit mass of a working substance that could be transformed into work in an engine under ideal conditions for a given amount of heat per unit mass furnished to the working substance.

Avogram: A unit of mass and weight equal to 1 gram divided by the Avogadro's number

Axial fan: A fan whose housing confines the gas flow to the direction along the rotating shaft at both the inlet and outlet.

Axial-flow pump: A pump having an axial flow or propeller-type impeller; used when maximum capacity and minimum head are de-sired. Also known as propeller pump.

Axial hydraulic thrust: In single stage and multistage pumps, the summation of unbalanced impeller forces acting in the axial direction.

Axial lead: A wire lead extending from the end along the axis of a resistor, capacitor, or other part.

Axial load: A force with its resultant passing through the centroid of a particular section and being perpendicular to the plane of the section.

Axial moment of inertia: For any object rotating about an axis, the sum of its component masses times the square of the distance to the axis.

Axial nozzle: An inlet or outlet connection installed in the head of a shell-and-tube exchanger and aligned normal to the plane in which the tube lies.

Axial rake: The angle between the face of a blade of a milling cutter or reamer and a line parallel to its axis of rotation.

Axis: A line about which a body rotates.

Axis of freedom: An axis in a gyro about which a gimbal offers a degree of freedom.

Axis of rotation: A straight line passing through the points of a rotating rigid body that remain stationary, while the other points of the body move in circles about the axis.

Axle: A supporting member that carries a wheel for power transmission, it allows the wheel to rotate freely on it.

Axle box: A bushing through which an axle passes in the hub of a wheel.

Axle ratio: In an automotive vehicle, the ratio of the speed in revolutions per minute of the drive shaft to that of the drive wheels.

Azimuth: In directional drilling, the direction of the face of the deviation tool with respect to magnetic north.

Back-draft damper: A damper with blades actuated by gravity, allowing air to pass through them in one way only.

Back edging: Cutting through a glazed ceramic tube by first chipping through the glaze around the outer and then chipping the tube itself.

Back flap hinge: A hinge having a flat plate or strap which is bolted to the face of a shutter or door. Also known as flap hinge.

Back flow: The flow of water or other liquids, mixtures, or substances into the distributing pipes of a potable supply of water from any other than its intended source.

Back flow connection: Any arrangement of pipes, plumbing fixtures, drains, and so forth, in which backflow can occur.

Back gearing: The technique of using gears on machine tools to obtain an increase in the number of speed changes that can be gotten with cone belt drives.

Backhoe: An excavator fitted with a hinged arm to which is rigidly attached a bucket that is drawn toward the machine in operation.

Backing pump: It is a vacuum pump. A vacuum pump is a device that draws gas molecules from a sealed volume in order to leave behind a partial vacuum.

Backlash: The amount by which the tooth space of a gear exceeds the tooth thickness of the mating gear along the pitch circles

Back pressure: Pressure due to a force that is operating in a direction opposite to that being considered, such as that of a fluid flow

Back rake: An angle on a single-point turning tool measured between the plane of the tool face and the reference plane.

Backsaw: A fine-tooth saw with its upper edge stiffened by a metal rib to ensure straight cuts.

Backward-bladed aerodynamic fan: A fan that consists of several streamlined blades climbed in a revolving casing.

Bailey meter: A flow meter consisting of a helical quarter turn vane which operate a counter to trace the total weight of granular substance.

Balanced draft: When the static pressure is equal to the atmospheric pressure, the system is referred to as balanced draft.

Balanced valve: A valve having the same fluid pressure in both the opening and closing directions.

Balance tool: A tool designed for taking the first cuts when the external face of a piece in a lathe is being machined.

Balance wheel: A wheel which directs or stabilizes the movement of a mechanism.

Baler: A machine which takes bulky quantities of raw or finished materials and binds them with rope or metal straps or wires into a large package.

Ball: In fine grinding, one of the crushing bodies used in a ball mill.

Ball-and-socket joint: A joint in which a member ending in a ball is joined to a member ending in a socket so that relative movement is permitted within a certain angle in all planes passing through a line.

Ball-and-trunnion joint: A joint in which a universal joint and a slip joint are combined in a solitary assembly.

Ball bearing: An antifriction bearing permitting free motion between moving and fixed parts by means of balls confined between outer and inner rings.

Ball-bearing hinge: A hinge which is equipped with ball bearings between the hinge knuckles in order to reduce friction

Ball bushing: A type of ball bearing that allows motion of the shaft in its axial direction.

Ball float: floating device, usually approximately spherical, which is used to operate a ball valve.

Ballhead: That part of the governor which contains flyweights whose force is balanced, at least in part, by the force of compression of a speeder spring.

Ballistic conditions: Conditions which affect the motion of a projectile in the bore and through the atmosphere, including muzzle velocity, weight of projectile

Ballistic curve: The curve described by the path of a bullet, a bomb, or other projectile as determined by the ballistic conditions

Ballistic deflection: The deflection of a missile due

to its ballistic characteristics

Ballistic density: A representation of the atmospheric density encountered by a projectile in flight.

Ballistic efficiency: The ability of a projectile to overcome the resistance of the air; depends chiefly on the weight, diameter, and shape of the projectile

Ballistic entry: Movement of a ballistic body from without to within a planetary atmosphere.

Ballistic limit: The minimum velocity at which a particular armor piercing projectile is expected to consistently and completely penetrate armor plate of given thickness and physical properties at a specified angle of obliquity.

Ballistic temperature: That temperature (in F) which, when regarded as a surface temperature and used in conjunction with the lapse rate of the standard artillery atmosphere, would produce the same effect on a projectile as the actual temperature distribution encountered by the projectile in flight.

Ballistic trajectory: The trajectory followed by a body being acted upon only by grav-

itational forces and resistance of the medium through which it passes.

Ballistic wave: An audible disturbance caused by compression of air ahead of a missile in flight.

Ball mill: A pulverizer that consists of a horizontal rotating cylinder, up to three diameters in length, containing a charge of tumbling or cascading steel balls,

Balloting: A tossing or bounding movement of a projectile, within the limits of the bore diameter, while moving through the bore under the influence of the propellant gases.

Ball screw: An element used to convert rotation to longitudinal motion, consisting of a threaded rod linked to a threaded nut by ball bearings constrained to roll in the space formed by the threads

Ball valve: A valve in which the fluid flow is regulated by a ball moving relative to a spherical socket as a result of fluid pressure and the weight of the ball.

Band brake: A brake in which the frictional force is applied by increasing the tension in a flexible band to tighten it around the drum.

Band clutch: A friction clutch in which a steel band, lined with fabric, contracts onto the clutch rim.

Band saw: A power operated wood working saw consisting basically of a flexible band of steel having teeth on one edge, running over two vertical pulleys, and operated under tension.

Band wheel: In a drilling operation, a large wheel that transmits power from the engine to the walking beam.

Bar: A unit of pressure equal to 10^5 pascals, or 10^5 newtons per square meter $(1Pas=1N/m^2)$

Bareboat charter: An agreement to charter a ship without its crew or stores; the fee for its use for a predetermined period of time is based on the price per ton of cargo handled

Bar linkage: A set of bars connected together at pivots by means of pins or equivalent devices; used to transmit power and information.

Baro dynamics: The mechanics of deep structures which may collapse under their own weight.

Barometric: An instrument for measuring atmospheric pressure, used especially in weather forecasting.

Barometric condenser: A contact condenser that uses a long, vertical pipe into which the condensate and cooling liquid flow to accomplish their removal by the pressure created at the lower end of the pipe

Baromil: The unit of length used in graduating a mercury barometer in the centimeter-gram-second system.

Barrel: 1. A container having a circular lateral cross section that is largest in the middle, and ends that are flat; often made of staves held together by hoops.

Bar turret lathe: A turret lathe in which the bar stock is slid through the headstock and collet on line with the turning axis of the lathe and held firmly by the closed collet.

Barycentric energy: The energy of a system in its center-of-mass frame.

Barye: The pressure unit of the centimeter-gram-second system of physical units; equal to 1 dyne per square centimeter (0.001 millibar). Also known as microbar.

Batch mixer: A machine which blend concrete or mortar in batches, as opposed to a continuous mixer.

Batch-type furnace: A furnace used for heat treatment of materials, with or without direct firing; loading and unloading operations are carried out through a single door or slot.

Bayonet-tube exchanger: A dual-tube apparatus with heating (or cooling) fluid flowing into the inner tube and out of the annular space between the inner and outer tubes; can be inserted into tanks or other process vessels to heat or cool the liquid contents.

Beaded tube end: The exposed portion of a rolled tube which is rounded back against the sheet in which the tube is rolled.

Beam-deflection amplifier: A jet-interaction fluidic device in which the direction of a supply jet is varied by flow from one or more control jets which are oriented at approximately 90 to the supply jet.

Bearing capacity: Load per unit area which can be securely supported by the ground.

Bearing strain: The deformation of bearing parts subjected to a load.

Bearing strength: The highest load that a column, wall, footing, or joint will sustain at failure, divided by the effective bearing area.

Beattie and Bridgman equation: An equation that relates the pressure, volume, and temperature of a real gas to the gas constant.

Bêche: A pneumatic forge hammer having an air drive ram and an air-compressing cylinder connected with the frame.

Bellows seal: A boiler seal in the form of a bellows which prevents leakage of air or gas.

Belt: Loop of flexible material between rotating shafts or pulleys.

Belt conveyor: A heavy duty conveyor consisting essentially of a head or drive pulley, a take-up pulley, a level or inclined endless belt made of canvas, rubber, or metal, and carrying and return idlers.

Belt drive: The transmission of power among shafts by means of a belt connecting pulleys on the shafts.

Belt feeder: A short belt conveyor used to transfer granulated or powdered solids from a storage or supply point to an end-use point; for example, from a bin hopper to a chemical reactor.

Belt guard: A cover designed to protect a belt as well as the pulleys it connects.

Belt sander: A portable sanding tool having a power-driven abrasive-coated continuous belt.

Belt shifter: A device with fingerlike projections used to shift a belt from one pulley to another or to replace a belt which has slipped off a pulley.

Belt slip: The difference in speed between the driving drum and belt conveyor.

Belt tightener: In a belt drive, a device that takes up the slack in a belt that has become stretched and permanently lengthened.

Bench lathe: A small engine or tool room lathe suitable for attachment to a workbench; bed length usually does not exceed 6 feet (1.8 meters) and work pieces are generally small.

Benchmark: A relatively permanent natural or artificial object bearing a marked point

Bench sander: A stationary power sander, usually mounted on a table or stand, which is equipped with a rotating abrasive disk or belt.

Bending: The forming of a metal part, by pressure, into a curved or angular shape, or the stretching or flanging of it along a curved path.

Bending brake: A press brake for making sharply angular linear bends in sheet metal.

Bending iron: A tool used to straighten or to expand flexible pipe, especially lead pipe.

Bending machine: A machine for bending a metal or wooden part by pressure. Also known as bender.

Bending moment: Algebraic sum of all moments located between a cross section and one end of a structural member;

Bending-moment diagram: A diagram showing the bending moment at every point along the length of a beam plotted as an ordinate.

Bending stress: An internal tensile or compressive longitudinal stress developed in a beam in response to curvature induced by an external load.

Bend wheel: A wheel used to interrupt and change the normal path of travel of the conveying or driving medium; most generally used to effect a change in direction of conveyor travel from inclined to horizontal or a similar change.

Bent-tube boiler: A water-tube steam boiler in which the tubes terminate in upper and lower steam-and-water drums. Also known as drum-type boiler.

Betti's method: A method of finding the solution of the equations of equilibrium of an elastic body whose surface displacements are specified; it uses the fact that the dilatation is a harmonic function to reduce the problem to the Dirichlet problem.

Betz momentum theory: A theory of windmill performance that considers the deceleration in the air traversing the windmill disk

Bevel gear: One of a pair of gears used to connect two shafts whose axes intersect.

Biaxial stress : The condition in which there are three mutually perpendicular principal stresses; two act in the same plane and one is zero.

Bicable tramway: A tramway consisting of two stationary cables on which the wheeled carriages travel, and an endless rope, which propels the carriages.

Bicycle: A human-powered land vehicle with two wheels, one behind the other, usually propelled by the action of the rider's feet on the pedals.

Bid: An estimate of costs for specified construction, equipment, or services proposed to a customer company by one or more supplier or contractor companies

Bin: An enclosed space, box, or frame for the storage of bulk substance.

Binary system: Any system containing two principal components.

Bionics: The study of systems, particularly electronic systems, which function after the manner of living systems.

Biot-Fourier equation: An equation for heat conduction

Bistable unit: A physical element that can be made to assume either of two stable states; a binary cell is an example.

Blackbody radiation: The emission of radiant energy which would take place from a blackbody at a fixed temperature; it takes place at a rate expressed by the Stefan-Boltzmann law, with a spectral energy distribution described by Planck's equation.

Blackbody temperature: The temperature of a blackbody that emits the same amount of heat radiation per unit area as a given object; measured by a total radiation pyrometer. Also known as brightness temperature.

Black smoke: A smoke that has many particulates in it from inefficient combustion; comes from burning fossil fuel, either coal or oil.

Black-surface enclosure: An enclosure for which the interior surfaces of the walls possess the radiation characteristics of a black-body.

Blacktop paver: A construction vehicle that spreads a specified thickness of bituminous mixture over a prepared surface.

Blank flange: A solid disk used to close off or seal a companion flange.

Blast: The setting off of a heavy explosive charge.

Blast burner: A burner in which a controlled burst of air or oxygen under pressure is supplied to the illuminating gas used. Also known as blast lamp.

Blast cleaning: Any cleaning process in which an abrasive is directed at high velocity toward the surface being cleaned, for example, sand blasting.

Blast heater: A heater that has a set of heat-transfer coils through which air is forced by a fan operating at a relatively high velocity.

Blast hole drilling: Drilling to produce a series of holes for placement of blasting charges.

Blast wall: A heavy wall used to isolate buildings or areas which contain highly combustible or explosive materials or to protect a building or area from blast damage when exposed to explosions.

Blears effect: The dependence of the signal from an ionization gage on the geometry of the system being measured when an organic vapor is present in the vacuum; the effect can falsify measurement results by up to an order of magnitude.

Bleed: To let a fluid, such as air or liquid oxygen, escape under controlled conditions from a pipe, tank, or the like through a valve or out-let.

Bleeder turbine: A multistage turbine where steam is extracted (bled) at pressures intermediate between throttle and exhaust, for process or feed water heating purposes.

Bleeding cycle: A steam cycle in which steam is drawn from the turbine at one or more stages and used to heat the feed water. Also known as regenerative cycle.

Bleed valve: A small-flow valve connected to a fluid process vessel or line for the purpose of bleeding off small quantities of contained fluid

Blind: A solid disk inserted at a pipe joint or union to prevent the flow of fluids through the pipe; used during maintenance and repair work as a safety precaution.

Blind drilling: Drilling in which the drilling fluid is not returned to the surface.

Blinding: A thin layer of lean concrete, fine gravel, or sand that is applied to a surface to smooth over voids in order to provide a cleaner, drier, or more durable finish. A layer of small rock chips applied over the surface of a freshly tarred road.

Blind joint: A joint which is not visible from any angle.

Blind nipple: A short piece of piping or tubing having one end closed off; commonly used in boiler construction.

Blind spot: An area on a filter screen where no filtering occurs. Also known as dead area.

Blink: A unit of time equal to 10^{-5} day or to 0.864 second.

Blistering: The appearance of enclosed or broken macroscopic cavities in a body or in a glaze or other coating during firing.

Block brake: A brake which consists of a block or shoe of wood bearing upon an iron or steel wheel.

Block diagram: A diagram in which the essential units of any system are drawn in the form of rectangles or blocks and their relation to each other is indicated by appropriate connecting lines.

Blocker-type forging: A type of forging for designs involving the use of large radii and draft angles, smooth contours, and generous allowances.

Block hole: A small hole drilled into a rock or boulder into which an anchor bolt or a small charge or explosive may be placed; used in quarries for breaking large blocks of stone or boulders.

Bloom: Fluorescence in lubricating oils or a cloudy surface on varnished or enameled surfaces.

Blotter: A disk of compressible material used between a grinding wheel and its flanges to avoid concentrated stress.

Blow by: Leaking of fluid between a cylinder and its piston during operation

Blower: A fan which operates where the resistance to gas flow is predominantly downstream of the fan.

Blowing: The introduction of compressed air near the bottom of a tank or other container in order to agitate the liquid therein.

Blowing pressure: Pressure of the air or other gases used to inflate the parison in blow molding.

Blow molding: A method of fabricating hollow plastic objects, such as bottles, by forcing a parison into a mold cavity and shaping by internal air pressure.

Blown glass: Glassware formed by blowing air into a ball of liquefied glass until it reaches the desired shape.

Blown tubing: A flexible thermoplastic film tube made by applying pressure inside a molten extruded plastic tube to expand it prior to cooling and winding flat onto rolls.

Blowoff valves: Valves in boiler piping which facilitate removal of solid matter present in the boiler water.

Blowpipe: A long, straight tube, used in glass blowing, on which molten glass is gathered and worked.

Blunging: The mixing or suspending of ceramic material in liquid by agitation, to form slip.

Body: The part of a drill which runs from the outer corners of the cutting lips to the shank or neck.

Body centrode: The path traced by the instantaneous center of a rotating body relative to the body.

Body force: An external force, such as gravity, which acts on all parts of a body.

Boiler: A water heater for generating steam.

Boiler air heater: A component of a steam-generating unit that transfers heat from the products of combustion after they have passed through the steam-generating and superheating sections to combustion air, which recycles heat to the furnace.

Boiler casing: The gas-tight structure surrounding the component parts of a steam generator.

Boiler circulation: Circulation of water and steam in a boiler, which is required to prevent overheating of the heat-absorbing surfaces; may be provided naturally by gravitational forces, mechanically by pumps, or by a combination of both methods.

Boiler code: A code, established by professional societies and administrative units, which contains the basic rules for the safe design, construction, and materials for steam-generating units, such as the American Society of Mechanical Engineers code.

Boiler controls: Either manual or automatic devices which maintain desired boiler operating conditions with respect to variables such as feedwater flow, firing rate, and steam temperature.

Boiler draft: The difference between atmospheric pressure and some lower pressure existing in the furnace or gas passages of a steam-generating unit.

Boiler economizer: A component of a steam-generating unit that transfers heat from the products of combustion after they have passed through the steam-generating and super-heating sections to the feedwater,

which it receives from the boiler feed pump and delivers to the steam-generating section of the boiler.

Boiler efficiency: The ratio of heat absorbed in steam to the heat supplied in fuel, usually measured in percent.

Boiler feedwater: Water supplied to a steam-generating unit.

Boiler feedwater regulation: Addition of water to the steam-generating unit at a rate commensurate with the removal of steam from the unit.

Boiler furnace: An enclosed space provided for the combustion of fuel to generate steam in a boiler. Also known as steam generating furnace.

Boiler heat balance: A means of accounting for the thermal energy entering a steam-generating system in terms of its ultimate useful heat absorption or thermal loss.

Boiler horsepower: A measurement of water evaporation rate; 1 boiler horsepower equals the evaporation per hour of $34^1/_2$ pounds (15.7 kilograms) of water at 212 F (100 C) into steam at 212 F.

Boiler hydrostatic test: A procedure that employs water under pressure, in a new boiler before use or in old equipment after major alterations and repairs, to test the boiler's ability to withstand about $1^1/_2$ times the design pressure.

Boiler layup: A significant length of time during which a boiler is inoperative in order to allow for repairs or preventive maintenance.

Boiler setting: The supporting steel and gastight enclosure for a steam generator.

Boiler storage: A steam-generating unit that, when out of service, may be stored wet (filled with water) or dry (filled with protective gas).

Boiler superheater: A boiler component, consisting of tubular elements, in which heat is added to high-pressure steam to increase its temperature and enthalpy.

Boiler trim: Piping or tubing close to or attached to a boiler for connecting controls, gages, or other instrumentation.

Boiler tube: One of the tubes in a boiler that carry water (water-tube boiler) to be heated by the high-temperature gaseous products of combustion or that carry combustion products

(fire-tube boiler) to heat the boiler water that surrounds them.

Boiler walls: The refractory walls of the boiler furnace, usually cooled by circulating water and capable of withstanding high temperatures and pressures.

Boiler water: Water in the steam-generating section of a boiler unit.

Boil-off: The vaporization of a liquid, such as liquid oxygen or liquid hydrogen, as its temperature reaches its boiling point under conditions of exposure, as in the tank of a rocket being readied for launch.

Bolometer: An instrument that measures the energy of electromagnetic radiation in certain wavelength regions by utilizing the change in resistance of a thin conductor caused by the heating effect of the radiation.

Bolster plate: A plate fixed on the bed of a power press to locate and support the die assembly.

Bolted joint: The assembly of two or more parts by a threaded bolt and nut or by a screw that passes through one member and threads into another.

Boltzmann engine: An ideal thermodynamic engine that utilizes blackbody radiation; used to derive the Stefan-Boltzmann law.

Bomb ballistics: The special branch of ballistics concerned with bombs dropped from aircraft.

Bomb test: A leak-testing technique in which the vessel to be tested is immersed in a pressurized fluid which will be driven through any leaks present.

Bonding strength: Structural effectiveness of adhesives, welds, solders, glues, or of the chemical bond formed between the metallic and ceramic components of a cermet, when subjected to stress loading, for example, shear, tension, or compression.

Bond's law: A statement that relates the work required for the crushing of solid materials (for example, rocks and ore) to the product size and surface area and the lengths

Bond strength: The amount of adhesion between bonded surfaces measured in terms of the stress required to separate a layer of material from the base to which it is bonded.

Boom: A row of joined floating timbers that extend across a river or enclose an area of water for the purpose of keeping saw logs together.

Boom cat: A tractor supporting a boom and used in laying pipe.

Boom dog: A ratchet device in-stalled on a crane to prevent the boom of the crane from being lowered but permitting it to be raised. Also known as boom ratchet.

Boom stop: A steel projection on a crane that will be struck by the boom if it is raised or lowered too great a distance.

Booster brake: An auxiliary air chamber, operated from the intake manifold vacuum, and connected to the regular brake pedal, so that less pedal pressure is required for braking.

Booster ejector: A nozzle-shaped apparatus from which a high-velocity jet of steam is discharged to produce a continuous-flow vacuum for process equipment.

Booster fan: A fan used to increase either the total pressure or the volume of flow.

Booster pump: A machine used to increase pressure in a water or compressed-air pipe.

Bootstrap: A technique or device designed to bring itself into a desired state by means of its own action.

Borer: An apparatus used to bore openings into the earth up to about 8 feet (2.4 meters) in diameter.

Boring bar: A rigid tool holder used to machine internal surfaces.

Boring machine: A machine tool designed to machine internal work such as cylinders, holes in castings, and dies; types are horizontal, vertical, jig, and single.

Boring mill: A boring machine tool used particularly for large workpieces; types are horizontal and vertical.

Bosch fuel injection pump: A pump in the fuel injection system of an internal combustion engine, whose pump plunger and barrel are a very close lapped fit to minimize leakage.

Bosch metering system: A system having a helical groove in the plunger which covers or uncovers openings in the barrel of the pump; most usually applied in diesel engine fuel-injection systems.

Bottom dead center (BDC): The bottom position of the piston of a vertical reciprocating engine.

Bottom flow: A molding apparatus that forms hollow plastic articles by injecting the blowing air at the bottom of the mold.

Bottom sampler: Any device used to obtain a sample from the bottom of a body of water.

Bounce table: A testing device which subjects devices and components to impacts such as might be encountered in accidental dropping.

Boundary friction: Friction between surfaces that are neither completely dry nor completely separated by a lubricant.

Boundary lubrication: A lubricating condition that is a combination of solid-to-solid surface contact and liquid-film shear.

Boundary monument: A material object placed on or near a boundary line to preserve and identify the location of the boundary line on the ground.

Bound vector: A vector whose line of application and point of application are both prescribed, in addition to its direction.

Boussinesq equation: A

relation used to calculate the influence of a concentrated load on the backfill behind a retaining wall.

Bowden cable: A wire made of spring steel which is enclosed in a helical casing and used to transmit longitudinal motions over distances, particularly around corners.

Bowl-mill pulverizer: A type of pulverizer which directly feeds a coal-fired furnace, in which springs press pivoted stationary rolls against a rotating bowl grinding ring, crushing the coal between them.

Brachistochrone: The curve along which a smooth-sliding particle, under the influence of gravity alone, will fall from one point to another in the minimum time (without friction).

Bradding: A distortion of a bit tooth caused by the appliance of too much weight, resulting the tooth to become dull so that its softer inner portion caves over the harder case area.

Braiding: Weaving fibers into a hollow cylindrical profile.

Brake: An element in a machine for applying friction to a moving surface to slow, down or bring it to rest.

Brake band: The constricting element of the band brake.

Brake block: A portion of the band brake lining, shaped to conform to the curvature of the band and attached to it with countersunk screws.

Brake drum: A rotating cylinder attached to a rotating part of machinery, which the brake band or brake shoe presses against.

Brake horsepower (BHP): The power available at the shaft of an engine. The power developed by an engine as measured by the force applied to a friction brake or by an absorption dynamometer applied to the shaft or flywheel.

Brake line: Ducts or hoses that connect the master cylinder and the wheel cylinders in a hydraulic brake system.

Brake lining: A casing, riveted or molded to the brake shoe or brake band, which presses against the rotating brake drum; made of either fabric or molded asbestos material.

Brake mean-effective pressure: It is applied to reciprocating piston machinery, the average pressure on the piston during the power stroke, derived from the measurement of brake power output.

Brake shoe: The renewable friction element of a shoe brake. Also known as shoe.

Brake thermal efficiency: The ratio of the brake power obtained from the engine to the fuel energy supplied to the engine.

Brayton cycle: A thermodynamic cycle consisting of two constant pressure processes interspersed with two Isentropic processes.

Brazed shank tool: A metal cutting tool made of a material different from the shank to which it is brazed.

Breaching: The space between the end of the tubing and the jacket of a hot-water or steam boiler.

Breaker cam: A rotating, engine driven device in the ignition system of an internal combustion engine which causes the breaker points to release, leading to a rapid fall in the primary current.

Break-even analysis: Determination of the break-even point.

Break-even point: The point at which a company neither create a profit nor un-

dergo a loss from the operations of the business, and at which total costs are equal to total sales volume.

Breaking load: The stress which, when steadily applied to a structural member, is just sufficient to break or burst it. Also known as ultimate load.

Breaking strength: The capability of a material to resist breaking or rupture from a tension force.

Breaking stress: The stress required to fracture a material whether by compression, tension, or shear.

Breather pipe: A pipe that opens into a container for airing, as in a crankcase or oil tank. Also known as crankcase breather.

Breeching: A pipe through which the products of combustion are transported from the furnace to the stack; usually applied in steam boilers.

Brennan monorail car: A type of car balanced on a single rail so that when the car starts to tip, a force automatically applied at the axle end is converted gyroscopically into a strong righting moment which forces the car back into a position of lateral equilibrium.

Bridge crane: A hoisting

machine in which the hoisting apparatus is carried by a bridge like structure spanning the area over which the crane operate.

Bridge trolley: Either of the wheeled attachments at the ends of the bridge of an overhead travelling crane, permitting the bridge to move backward and forward on elevated path.

Bridge vibration: Mechanical vibration of a bridge superstructure due to innate and human produced excitations.

Bridgewall: The wall in a furnace over which the products of combustion flow.

Brinell number: A hardness rating obtained from the Brinell hardness test, expressed in kilo-gram force per square millimeter. P = applied load in kilogram-force (kgf): D = diameter of indenter (mm): d = diameter of indentation (mm).

British imperial pound: The British standard of mass, used in British imperial and United States.

Brittle temperature: The temperature point under which a material, especially metal, is brittle; that is, the

critical normal stress for fracture is reached prior to the critical shear stress for plastic deformation.

Brush rake: Brush Rakes are utilized in typical land clearing applications. A device with heavy-duty tines that is fixed to the front of a tractor or other prime mover for use in land clearing.

Bucket conveyor: An incessant bulk conveyor constructed of a series of buckets attached to one or two strands of chain or in some instances to a belt. Also called bucket carrier.

Bucket dredge: A bucket dredger is equipped with a bucket dredge. which is used to pick sediment by mechanical means.

Bucket elevator: A bucket conveyor operating on a steep incline or vertical path. Which is also known as elevating conveyor.

Bucket loader: A form of portable, self-feeding, inclined bucket elevator for loading largeness materials into cars, trucks, or other conveyors.

Bucket temperature: The surface temperature of ocean water as measured by a bucket thermometer.

Bulking factor: The bulking factor is the ratio or percentage of the volume change of excavated material to the volume of the original in situ volume before excavation.

Bulk-handling machine: Any of a diversified set of materials handling machines designed for handling unpackaged, divided materials.

Bulk modulus of elasticity: Bulk modulus of elasticity is simply the ratio of the pressure to the corresponding volumetric strain. Also known as bulk modulus; compression modulus; hydrostatic modulus; modulus of compression; modulus of volume elasticity.

Bulk strain: Bulk Strain may also be known as volume strain. The ratio of the change in the volume of a body that occurs when the body is placed under pressure, to the original volume of the body.

Bulk strength: The strength per unit volume of a solid. (strength/volume)

Bulk transport: Conveying, hoisting, or elevating systems for movement of solids such as grain, sand, gravel, coal, or wood chips.

Bulldozer: It is a large, motorized machine. A wheeled or crawler tractor equipped with a reinforced, curved steel plate mounted in front, perpendicular to the ground, for pushing excavated materials.

Bull wheel: The major wheel or gear of a machine, which is generally the largest and strongest.

Burnside boring machine: A machine for boring in all types of ground with the feature of controlling water instantly if it is tapped.

Bursting strength: A measure of the capability of a material to withstand pressure without rupture.

Burst pressure: The maximum inside pressure that a pressure vessel can safely with-stand.

Burton: A small hoisting tackle by two blocks, generally a single block and a double block, with a hook block in the running part of the rope.

Bushel: A bushel is an imperial and US customary unit of volume based upon an earlier measure of dry capacity. The old bushel is equal to 2 kennings, 4 pecks, or 8 dry gallons, and was used mostly for agricultural products.

Bush hammer: A hand-held or power-driven hammer that has a serrated face containing pyramid shaped points and is used to dress a concrete or stone surface.

Butterworth head: A mechanical hose head with revolving nozzles; which is used to wash down shipboard storage tanks.

Bypass valve: The valve that serve to control pressure in a system by diverting a portion of the flow.

By-product: A product from a manufacturing process that is not considered as the primary material.

Cab: In a locomotive, truck, tractor, or hoisting apparatus, a partition for the operator.

Cable railway: An inclined path on which rail cars travel, with the cars fixed to an endless steel-wire rope at equal spaces.

Cableway: A transporting arrangement consisting of a cable extended between two or more points on which cars are propelled to transport bulk materials for construction operations.

Cage: A frame for upholding uniform separation between the balls or rollers in a bearing. Also known as separator.

Cage mill: Pulverizer used to disintegrate clay, press cake, asbestos, packing-house by-products, and various tough, gummy, high-moisture-content or low-melting-point materials.

Callendar's equation: An equation of state for steam whose temperature is well above the boiling point

Calorie: Abbreviated cal; often designated c. A unit of heat energy, equal to 4.1868 joules. calorific value: Quantity of heat liberated on the complete combustion of a unit weight or unit volume of fuel.

Cam: A plate or cylinder which communicates motion to a follower by means of its edge or a groove cut in its surface. Cams are mechanical devices used to convert the rotation of a shaft into simple or complex reciprocating linear motion.

Cam acceleration: The acceleration of the cam follower. Axis of cam and follower offset with a uniform acceleration.

Cam cutter: A semiautomatic or automatic machine that produces the cam contour by swinging the work as it revolves; uses a master cam in contact with a roller.

Cam engine: A piston engine in which a cam-and-roller mechanism seems to convert reciprocating motion into rotary motion

Cam follower: The Cam Follower is a compact and highly rigid bearing with a shaft. It contains needle rollers and is used as a guide roller for cam mechanisms or straight motion.

Cam mechanism: A mechanical linkage whose purpose is to produce, by means of a contoured cam surface, a prescribed motion of the output link.

Cam nose: The high point of a cam, which in a reciprocating engine holds valves open or closed.

Cam pawl: A pawl which prevents a wheel from turning in one direction by a wedging action, while permitting it to rotate in the other direction.

Camshaft: A rotating shaft to which a cam is attached.

Canned pump: A watertight pump that can operate under water. The Canned Motor Pump is composed of a traditional centrifugal hydraulic, connected to a special motor through a one piece assembly.

Cantilever spring: A flat spring supported at one end and holding a load at or near the other end.

Cantilever vibration: Transverse oscillatory motion of a body fixed at one end.

Canting: Displacing the free end of a beam which is fixed at one end by subjecting it to a sideways force which is just short of that required to cause fracture.

Cape foot: A unit of length capillary tube : A tube sufficiently fine so that capillary attraction of a liquid into the tube is significant.

Carathéodory's principle: An expression of the second law of thermodynamics which says that in the neighborhood of any equilibrium state of a system, there are states which are not accessible by a reversible or irreversible adiabatic process.

Carbon knock: Premature ignition resulting in knocking or pinging in an internal combustion engine caused when the accumulation of carbon produces overheating in the cylinder.

Cardan motion: The straight-line path followed by a moving centrode in a four-bar centrode linkage.

Cardan shaft: A shaft with a universal joint at its end to accommodate a varying shaft angle. A shaft that has a universal joint at one or both ends enabling it to rotate freely when in varying angular relation to another shaft or shafts to which it is joined.

Car dump: Any one of several devices for unloading industrial or railroad cars by rotating or tilting the car.

Cargo winch: A motor-driven hoisting machine for cargo having a drum around which a chain or rope winds as the load is lifted.

Carnot cycle: A hypothetical cycle consisting of four reversible processes in succession: an isothermal expansion and heat addition, an isentropic expansion, an isothermal compression and heat rejection process, and an isentropic compression.

Carnot efficiency: The efficiency of a Carnot engine Carnot efficiency is the maximum efficiency that a heat engine may have operating between the two temperatures

Carnot engine: An ideal, frictionless engine which operates in a Carnot cycle. Carnot engine is a theoretical thermodynamic cycle proposed by Leonard Carnot. It gives the estimate of the maximum possible efficiency that a heat engine

Carnot number: A property of two heat sinks, equal to the Carnot efficiency of an engine operating between them.

Carnot's theorem: The theorem that all Carnot engines operating between two given temperatures have the same efficiency, and no cyclic heat engine operating between two given temperatures is more efficient than a Carnot engine.

Carousel: A rotating transport system that trans-

fers and presents workpieces for loading and unloading by a robot or other ma-chine.

Carriage stop: A device added to the outer way of a lathe bed for accurately spacing grooves, turning multiple diameters and lengths, and cutting off pieces of specified thickness.

Carrier: Any machine for transporting materials or people.

Car shaker: A device consisting of a heavy yoke on an open-top car's sides that actively vibrates and rapidly discharges a load, such as coal, gravel, or sand, when an unbalanced pulley attached to the yoke is rotated fast.

Cascading drain: A flow of water into the closed shell of a feedwater heater from a water source maintained at a higher pressure. Cascade drain is one of the hydraulic structures that are widely used in slope. condition area. The functions of cascade drain are to control and convey flows from. surface runoff at upstream to downstream.

Causality: In classical mechanics, the principle that the specification of the dynamical variables of a system at a given time, and of the external forces acting on the system, completely determines the values of dynamical variables at later times.

Cavity radiator: A heated enclosure with a small opening which allows some radiation to escape or enter; the escaping radiation approximates that of a blackbody.

Celo: A unit of acceleration equal to the acceleration of a body whose velocity changes uniformly by 1 foot (0.3048 meter) per second in 1 second.

Celsius temperature scale: Temperature scale in which the temperature c in degrees Celsius (°C) is related to the temperature T_k in kelvins.

Cement gun: A mechanical device for the application of cement, in the form of gunite, to the walls or roofs of mine openings or building walls. Also called gunite gun

Cement mill: A mill for grinding rock to a powder for cement. A cement mill is the equipment used to grind the hard, nodular clinker from the cement kiln into the fine grey powder that is cement.

Cement pump: It is known as a boom concrete pump because it uses a remote-controlled articulating robotic arm (called a boom) to place concrete accurately.

Cement valve: A ball-, flapper-, or clack-type valve placed at the bottom of a string of casing, through which cement is pumped, so that when pumping ceases, the valve closes and prevents return of cement into the casing.

Centering machine: Centering machine is a turning machine that can machine both ends of a piece of bar stock simultaneously.

Centerless grinder: Centerless grinding is a machining process that uses abrasive cutting to remove material from a workpiece.

Center of attraction: A point toward which a force on a body or particle (such as gravitational or electrostatic force) is always directed; the magnitude of the force depends only on the distance of the body or particle from this point.

Center of buoyancy: The point through which acts the resultant force exerted on a body by a static fluid in which it is submerged or floating; located at the centroid of displaced volume.

Center of force: The point toward or from which a central force acts.

Center of gravity: Centre of gravity is the point where the mass of the body is concentrated. The centre of gravity (COG) of the human body is a hypothetical point around which the force of gravity appears to act. It is point at which the combined mass of the body appears to be concentrated

Center of mass :That point of a material body or system of bodies which moves as though the system's total mass existed at the point and all external forces were applied at the point. Also known as center of inertia; centroid.

Center-of-mass coordinate system: A reference frame which moves with the velocity of the center of mass, so that the center of mass is at rest in this system, and the total momentum of the system is zero. Also known as center of momentum coordinate system.

Center of oscillation: Point in a physical pendulum, on the line through the point of suspension and the center of mass, which moves as if all the mass of the pendulum were concentrated there.

Center of percussion: If a rigid body, free to move in a plane, is struck a blow at a point O, and the line of force is perpendicular to the line from O to the center of mass, then the initial motion of the body is a rotation about the center of percussion relative to O; it can be shown to coincide with the center of oscillation relative to O.

Center of suspension: The intersection of the axis of rotation of a pendulum with a plane perpendicular to the axis that passes through the center of mass.

Center of twist: A point on a line parallel to the axis of a beam through which any transverse force must be applied to avoid twisting of the section. Also known as shear center.

Centiare: The centiare is a unit of area used in the metric system, and is a synonym for 1 square meter (m2). The abbreviation for centiare is ca.

Centibar: A unit of pressure equal to 0.01 bar or to 1000 pascals.

Centigrade heat unit: A unit of heat energy, equal to 0.01 of the quantity of heat needed to raise 1 pound of air-free water from 0 to 100 C at a constant pressure of 1 standard atmosphere; equal to

1900.44 joules

Centigram: Unit of mass equal to 0.01 gram

Centilitre: A unit of volume equal to 0.01 liter or to 10^{-5} cubic meter.

Centimeter: A centimeter is a metric unit of measurement used for measuring the length of an object. It is written as cm.

Central force: Central force is a force (possibly negative) that points from the particle directly towards a fixed point in space, the center, and whose magnitude only depends on the distance of the object to the center.

Central gear: The gear on the central axis of a planetary gear train, about which a pinion rotates. Also known as sun gear.

Central orbit: The path followed by a body moving under the action of a central force.

Centrifugal: Acting or moving in a direction away from the axis of rotation or the center of a circle along which a body is moving.

Centrifugal brake: A safety device on a hoist drum that applies the brake if the drum speed is greater than a set limit.

Centrifugal clarification: The removal of solids from a liquid by centrifugal action which decreases the settling time of the particles from hours to minutes.

Centrifugal classification: A type of centrifugal clarification purposely designed to settle out only the large particles (rather than all particles) in a liquid by reducing the centrifuging time.

Centrifugal classifier: A machine that separates particles into size groups by centrifugal force.

Centrifugal clutch: A clutch operated by centrifugal force from the speed of rotation of a shaft, as when heavy expanding friction shoes act on the internal surface of a rim clutch, or a flyball-type mechanism is used to activate clutching surfaces on cones and disks.

Centrifugal collector: Device used to separate particulate matter of 0.1-1000 micrometers from an airstream; some types are simple cyclones, high-efficiency cyclones, and impellers.

Centrifugal compressor: A machine in which a gas or vapor is compressed by radial acceleration in an impeller with a surrounding casing,

Centrifugal fan: A machine for moving a gas, such as air, by accelerating it radially outward in an impeller to a surrounding casing, generally of scroll shape.

Centrifugal force: An outward pseudo-force, in a reference frame that is rotating with respect to an inertial reference frame, which is equal and opposite to the centripetal force that must act on a particle stationary in the rotating frame.

Centrifugal governor: Centrifugal governor is a specific type of governor with a feedback system that controls the speed of an engine by regulating the flow of fuel or working.

Centrifugal moment: The product of the magnitude of centrifugal force acting on a body and the distance to the center of rotation

Centrifugal pump: A machine for moving a liquid, such as water, by accelerating it radially outward in an impeller to a surrounding volute casing.

Centrifugal separation: The separation of two immiscible liquids in a centrifuge within a much shorter period of time than could be accomplished solely by gravity.

Centrifugal switch: A switch opened or closed by centrifugal force; used on some induction motors to open the starting winding when the motor has almost reached synchronous speed.

Centrifugal tachometer: An instrument which measures the instantaneous angular speed of a shaft by measuring the centrifugal force on a mass rotating with it.

Centripetal: Acting or moving in a direction toward the axis of rotation or the center of a circle along which a body is moving.

Centripetal acceleration: The radial component of the acceleration of a particle or object moving around a circle, which can be shown to be directed toward the center of the circle. Also known as radial acceleration

Centripetal force: The radial force required to keep a particle or object moving in a circular path, which can be shown to be directed toward the center of the circle.

Centrobaric: Pertaining to the center of gravity, or to

some method of locating it.

Centrode: The path traced by the instantaneous center of a plane figure when it undergoes plane motion.

Chain conveyor: A machine for moving materials that carries the product on one or two endless linked chains with crossbars; allows smaller parts to be added as the work passes.

Chain drive: A flexible device for power transmission, hoisting, or conveying, consisting of an endless chain whose links mesh with toothed wheels fastened to the driving and driven shafts.

Chain gear: A gear that transmits motion from one wheel to another by means of a chain.

Chain grate stoker: A wide, endless chain used to feed, carry, and burn a noncoking coal in a furnace, control the air for combustion, and discharge the ash.

Chain pump: A pump containing an endless chain that is fitted at intervals with disks and moves through a pipe and raises sludge

Chain saw: A gasoline-powered saw for felling and bucking timber, operated by one person; has cutting teeth inserted in a sprocket chain that moves rapidly around the edge of an oval-shaped blade.

Chamfering: Machining operations to produce a beveled edge. Also known as beveling.

Change gear: A gear used to change the speed of a driven shaft while the speed of the driving remains constant.

Characteristic length: A convenient reference length (usually constant) of a given configuration, such as overall length of an aircraft, the maximum diameter or radius of a body of revolution, or a chord or span of a lifting surface.

Chase ring: In hobbing, the ring which restrains the blank from spreading during hob sinking.

Check valve: A device for automatically limiting flow in a piping system to a single direction. Also known as non-return valve.

Cherry picker: Any of several small travelling cranes, especially one used to hoist a passenger on the end of a boom.

Chicago boom: A hoisting device that is supported on the structure being erected.

Chimney core: The inner section of a double-walled chimney which is separated from the outer section by an air space.

Chladni's figures: Figures produced by sprinkling sand or similar material on a horizontal plate and then vibrating the plate while holding it rigid at its center or along its periphery; indicate the nodal lines of vibration.

Choke valve: A valve which sup- plies the higher suction necessary to give the excess fuel feed required for starting a cold internal combustion engine. Also known as choke.

Chopping bit: A steel bit with a chisel-shaped cutting edge, attached to a string of drill rods to break up, by impact, boulders, hardpan, and a lost core in a drill hole. Also known as chisel bit.

Chucking: The grasping of an outsize workpiece in a chuck or jawed device in a lathe.

Chucking machine: A lathe or grinder in which the outsize workpiece is grasped in a chuck or jawed device.

Churn shot drill: A boring rig with both churn and shot drillings. A check valve is a device that only allows the flow of fluids in one direction. They have two ports, one as an inlet for the media and one as the output

Circle shear: These machines cut circles and rings from a square metal piece. For increased power and efficiency.

Circular inch: The area of a circle 1 inch (25.4 millimeters) in diameter. The circular inch is a unit of area in the Imperial and US.

Circular mil: A unit equal to the area of a circle whose diameter is 1 mil (0.001 inch); used chiefly in specifying cross-sectional areas of round conductors. Abbreviated cir mil.

Circular motion: Circular motion is a movement of an object along the circumference of a circle or rotation along a circular path

Circular saw: A circular saw is a power-saw using a toothed or abrasive disc or blade to cut different materials using a rotary motion spinning around an arbor.

Circular velocity: At any specific distance from the primary, the orbital velocity required to maintain a constant-radius orbit.

Clamping coupling: A

coupling with a split cylindrical element which clamps the shaft ends together by direct compression, through bolts or rings, and by the wedge action of conical sections; not considered a permanent part of the shaft.

Clamshell bucket: A two-sided bucket used in a type of excavator to dig in a vertical direction; the bucket is dropped while its leaves are open and digs as they close. Also known as clamshell grab.

Clapeyron's theorem: The theorem that the strain energy of a deformed body is equal to one-half the sum over three perpendicular directions of the displacement component times the corresponding force component, including deforming loads and body forces, but not the six constraining forces required to hold the body in equilibrium.

Clapper box: A hinged device that permits a reciprocating cutting tool (as in a planer or shaper) to clear the work on the return stroke.

Clarifying centrifuge: A device that clears liquid of foreign matter by centrifugation classical mechanics: Mechanics based on Newton's laws of motion.

Classifier: Any apparatus for separating mixtures of materials into their constituents according to size and density.

Clausius: A unit of entropy equal to the increase in entropy associated with the absorption of 1000 international table calories of heat at a temperature of 1 K, or to 4186.8 joules per kelvin.

Clausius equation: An equation of state in reference to gases which applies a correction to the van der Waals equation:

Clausius inequality: The principle that for any system executing a cyclical process, the integral over the cycle of the infinitesimal amount of heat transferred to the system divided by its temperature is equal to or less than zero. Also known as Clausius theorem; inequality of Clausius.

Clausius law: It is impossible to design a device which works on a cycle and produce no other effect other than heat transfer from a cold body to a hot body."

Clausius number: A dimensionless number used in the study of heat conduction in forced fluid flow, equal to V3L /k T, where V is the fluid velocity, is its density, L is a characteristic dimension, k is the thermal conductivity, and T is the temperature difference.

Claw clutch: A mechanical clutch in which jaws or claws interlock when pushed together. It is also called as positive clutch.

Claw coupling: A loose coupling having projections or claws cast on each face which engage in corresponding notches in the opposite faces; used in situations in which shafts require instant connection.

Cleaning turbine: A tool for cleaning the interior surfaces of heat exchangers and boiler tubes; consists of a drive motor, a flexible drive cable or hose, and a head that is an arrangement of blades,

Clearance angle: The angle between a plane containing the end surface of a cutting tool and a plane passing through the cutting edge in the direction of cutting motion

Clearance volume: The volume remaining between piston and cylinder when the piston is at top dead center. close-coupled pump: Pump with built-in electric motor (sometimes a steam turbine), with the motor drive and

pump impeller on the same shaft.

Closed-belt conveyor: Solids conveying device with zipperlike teeth that mesh to form a closed tube wrapped snugly around the conveyed material; used with fragile materials.

Closed cycle: A thermodynamic cycle in which the thermodynamic fluid does not enter or leave the system, but is used over and over again.

Closed-cycle turbine: A gas turbine in which essentially all the working medium is continuously recycled, and heat is transferred through the walls of a closed heater to the cycle.

Closed fireroom system: A fireroom system in which combustion air is supplied via forced draft resulting from positive air pressure in the fireroom.

Closed nozzle : A fuel nozzle having a built-in valve interposed between the fuel supply and combustion chamber.

Closed pair: A pair of bodies that are subject to constraints which prevent any relative motion between them.

Close-off rating: The maximum allowable pressure drop

to which a valve can be sub-jected at commercial shutoff. 2. The maximum allowable pressure drop between the out-let of a three-way valve and either of the two inlets, or between the inlet and ei-ther of the two outlets.

Closing line: The vector required to complete a poly-gon consisting of a set of vec-tors whose sum is zero

Closing pressure: The amount of static inlet pres-sure in a safety relief valve when the valve disk has a zero lift above the seat.

Clusec: A unit of power used to measure the power of evacuation of a vacuum pump, equal to the power associated with a leak rate of 1 centiliter per second at a pressure of 1 millitorr, or to approximately $1.333221 \cdot 10^{-6}$ watt.

Clutch: A mechanism for connecting and disconnecting an engine and the transmis-sion system in a vehicle, or the working parts of any machine.

Coaxial: Sharing the same axes. Mounted on independ-ent concentric shafts.

Coefficient of compress-ibility: The decrease in vol-ume per unit volume of a sub-stance resulting from a unit increase in pressure; it is the reciprocal of the bulk modulus.

Coefficient of cubical ex-pansion: The increment in volume of a unit volume of solid, liquid, or gas for a rise of temperature of 1 at con-stant pressure.

Coefficient of friction: The ratio of the frictional force between two bodies in contact, parallel to the sur-face of contact, to the force, normal to the surface of con-tact, with which the bodies press against each other. Also known as friction coeffi-cient.

Coefficient of kinetic friction: The ratio of the frictional force, parallel to the surface of contact, that op-poses the motion of a body which is sliding or rolling over another, to the force, normal to the surface of contact, with which the bodies press against each other.

Coefficient of linear ex-pansion: The increment of length of a solid in a unit of length for a rise in tempera-ture of 1 at constant pres-sure. Also known as linear expansivity.

Coefficient of perfor-mance: In a refrigeration cycle, the ratio of the heat energy extracted by the heat engine at the low temperature to the work supplied to operate the cycle; when used as a heating device, it is the ratio of the heat delivered in the high-temperature coils to the work supplied.

Coefficient of restitution: The constant e, which is the ratio of the relative velocity of two elastic spheres after direct impact to that before impact; e can vary from 0 to 1, with 1 equivalent to an elastic collision and 0 equivalent to a perfectly elastic collision.

Cog belt: A cog belt is a high-performance belt used in applications where slippage or failure must be avoided.

Cogeneration: The simultaneous on-site generation of electric energy and process steam or heat from the same plant.

Coke knocker: A mechanical device used to break loose coke within a drum or tower

Cold plate: An aluminum or other plate containing internal tubing through which a liquid coolant is forced, to absorb heat transferred to the plate by transistors and other components mounted on it. Also known as liquid-cooled dissipator.

Cold saw: A cold saw is a circular saw designed to cut metal which uses a toothed blade to transfer the heat generated by cutting to the chips created by the saw blade, allowing both the blade and material being cut to remain cool.

Collapse properties: Strength and dimensional attributes of piping, tubing, or process vessels, related to the ability to resist collapse from exterior pressure or internal vacuum.

Collapsing pressure: The minimum external pressure which causes a thin-walled body or structure to collapse.

Collar bearing: A thrust bearing having a suitably formed face or faces that resist the axial pressure of one or more collars on a rotating shaft.

Colloid mill: A grinding mill for the making of very fine dispersions of liquids or solids by breaking down particles in an emulsion or paste.

Column crane: A jib crane whose boom pivots about a post attached to a building column.

Column drill: Column drill presses are mostly used in industrial applications, but also versatile solutions for workshops. This powerful universal drill press is ideally equipped for large bore diameters and high-precision thread cutting tasks

Combination saw: A combination circular saw blade is a design compromise between a rip cutting and crosscutting blade. It is designed to do both, by making design tradeoffs

Combined flexure: The flexure of a beam under a combination of transverse and longitudinal loads.

Combined stresses: The combined stresses can be obtained by superposition of the bending stresses and the axial stresses

Combustion-chamber volume: The volume of the combustion chamber when the piston is at top dead center.

Combustion engine: An engine that operates by the energy of combustion of a fuel.

Combustion engineering: The design of combustion furnaces for a given performance and thermal efficiency, involving study of the heat liberated in the combustion

process, the amount of heat absorbed by heat elements, and heat-transfer rates.

Combustor: The combustion chamber together with burners, igniters, and injection devices in a gas turbine or jet engine

Comfort temperature: Any one of the indexes in which air temperatures have been adjusted to represent human comfort or discomfort under prevailing conditions of temperature, humidity, radiation, and wind.

Comminution: Comminution is the reduction of solid materials from one average particle size to a smaller average particle size, by crushing, grinding, cutting

Comminutor: A machine that breaks up solids.

Comparator method: A method of determining the coefficient of linear expansion of a substance in which one measures the distance that each of two traveling microscopes must be moved in order to remain centered on scratches on a rod-shaped specimen when the temperature of the specimen is raised by a measured amount.

Compartment mill: A multisection pulverizing device divided by perforated partitions, with preliminary grinding at one end in a short ball-mill operation, and finish grinding at the discharge end in a longer tube-mill operation.

Compatibility conditions: A set of six differential relations between the strain components of an elastic solid which must be satisfied in order for these components to correspond to a continuous and single-valued displacement of the solid.

Compliance: The displacement of a linear mechanical system under a unit force.

Compliance constant: Any one of the coefficients of the relations in the generalized Hooke's law used to express strain components as linear functions of the stress components. Also known as elastic constant.

Composition: The determination of a force whose effect is the same as that of two or more given forces acting simultaneously; all forces are considered acting at the same point.

Composition-of-velocities law: A law relating the velocities of an object in two references frames which are moving relative to each other with a specified velocity.

compound engine: A multi cylinder-type displacement engine, using steam, air, or hot gas, where expansion proceeds successively (sequentially).

Compounding: The series placing of cylinders in an engine (such as steam) for greater ratios of expansion and consequent improved engine economy.

Compound lever: A train of levers in which motion or force is transmitted from the arm of one lever to that of the next.

Compound rest: A principal component of a lathe consisting of a base and an upper part dovetailed together; the base is graduated in degrees and can be swiveled to any angle; the upper part includes the tool post and tool holder.

Compressadensity function: A function used in the acoustic levitation technique to determine either the density or the adiabatic compressibility of a submicroliter droplet suspended in another liquid, if the other property is known

Compressed air: Air whose

density is increased by subjecting it to a pressure greater than atmospheric pressure. Compressing the air makes the molecules move more rapidly, which increases the temperature. This phenomenon is called "heat of compression.

Compressed-air power: The power delivered by the pressure of compressed air as it expands, utilized in tools such as drills, in hoists, grinders, riveters, diggers, pile drivers, motors, locomotives, and in mine ventilating systems.

Compressibility: The property of a substance capable of being reduced in volume by application of pressure; quantitively, the reciprocal of the bulk modulus.

Compressibility factor: The product of the pressure and the volume of a gas, divided by the product of the temperature of the gas and the gas constant; this factor may be inserted in the ideal gas law to take into account the departure of true gases from ideal gas behavior. Also known as deviation factor; gas-deviation factor; super-compressibility factor.

Compression coupling: A means of connecting two perfectly aligned shafts in which a slotted tapered sleeve is placed over the junction and two flanges are drawn over the sleeve so that they automatically center the shafts and provide sufficient contact pressure to transmit medium loads.

Compression ignition: In the compression ignition cycle the air is compressed and the fuel is injected into the compressed air at a temperature sufficiently high to spontaneously .

Compression refrigeration: The cooling of a gaseous refrigerant by first compressing it to liquid form.

Compression release: Release of compressed gas resulting from incomplete closure of intake or exhaust valves.

Compression ring: A ring located at the upper part of a piston to hold the burning fuel charge above the piston in the combustion chamber, thus preventing blowby.

Compression strength: Property of a material to resist rupture under compression

Compression stroke: The phase of a positive displacement engine or compressor in which the motion of the piston compresses the fluid trapped in the cylinder

Compressive strength: Compressive strength or compression strength is the capacity of a material or structure to withstand loads tending to reduce size.

Compressive stress: A stress which causes an elastic body to shorten in the direction of the applied force.

Compressor blade: The vane components of a centrifugal or axial-flow, air or gas compressor.

Compressor station: A permanent facility which increases the pressure on gas to move it in transmission lines or into storage

Compressor valve: A valve in a compressor, usually automatic, which operates by pressure difference on the two sides of a movable, single-loaded member and which has no mechanical linkage with the moving parts of the compressor mechanism.

Concentrated load: A force that is negligible because of a small contact area; a beam supported on a girder represents a concentrated load on the girder.

Concrete mixer: A concrete mixer is a device that homogeneously combines cement, aggregate such as sand or gravel, and water to form concrete.

Concrete pump: A device which drives concrete to the placing position through a pipeline of 6-inch (15-centimeter) diameter or more, using a special type of reciprocating pump

Concrete vibrator: Vibrating device used to achieve proper consolidation of concrete; the three types are internal, surface, and form vibrators.

Condensate strainer: Strainer is device that captures solids in liquids, gases or steam. Strainer protects equipment from harmful influences.

Condensate well: A chamber into which condensed vapor falls for convenient accumulation prior to removal.

Condenser: A heat-transfer device that reduces a thermo-dy-namic fluid from its vapor phase to its liquid phase, such as in a vapor-compression refrigera-tion plant or in a condensing steam power plant.

Condenser tubes: Metal tubes used in a heat-transfer device, with condenser vapor as the heat source and flowing liquid such as water as the receiver.

Condensing engine: A steam engine in which the steam exhausts from the cylinder to a vacuum space, where the steam is liquefied.

Conditionally periodic motion: Motion of a system in which each of the coordinates undergoes simple periodic motion, but the associated frequencies are not all rational fractions of each other so that the complete motion is not simply periodic.

Cone brake: A friction brake in which the frictional surfaces are cone-shaped.

Cone classifier: Inverted-cone device for the separation of heavy particulates (such as sand, ore, or other mineral matter) from a liquid stream; feed enters the top of the cone, heavy particles settle to the bottom where they can be withdrawn, and liquid overflows the top edge, carrying the smaller particles or those of lower gravity over the rim; used in the mining and chemical industries.

Cone clutch: A cone clutch serves the same purpose as a disk or plate clutch. However, instead of mating two spinning disks, the cone clutch uses two conical surfaces to transmit torque by friction.

Cone crusher: A machine that reduces the size of materials such as rock by crushing in the tapered space between a truncated revolving cone and an outer chamber.

Cone of friction: A cone in which the resultant force exerted by one flat horizontal surface on another must be located when both surfaces are at rest, as determined by the coefficient of static friction.

Cone rock bit: A rotary drill with two hardened knurled cones which cut the rock as they roll. Also known as roller bit.

Congruent melting point: A point on a temperature composition plot of a non-stoichiometric compound at which the one solid phase and one liquid phase are adjacent.

Conical ball mill: A cone-shaped tumbling pulverizer in which the steel balls are classified, with the larger balls at the feed end where larger lumps are crushed, and the smaller balls at the discharge end where the material is finer.

Conical bearing: An antifriction bearing employing tapered rollers

Conical pendulum: A weight suspended from a cord or light rod and made to rotate in a horizontal circle about a vertical axis with a constant angular velocity.

Conical refiner: In paper manufacturing, a cone-shaped continuous refiner having two sets of bars mounted on the rotating plug and fixed shell for beating unmodified cellulose fibers.

Connecting rod: Any straight link that transmits motion or power from one linkage to another within a mechanism, especially linear to rotary motion, as in a reciprocating engine or compressor.

Conservation of angular momentum: The principle that, when a physical system is subject only to internal forces that bodies in the system exert on each other, the total angular momentum of the system remains constant, provided that both spin and orbital angular momentum are taken into account.

Conservation of areas: A principle governing the motion of a body moving under the action of a central force, according to which a line joining the body with the center of force sweeps out equal areas in equal times.

Conservation of momentum: The principle that, when a system of masses is subject only to internal forces that masses of the system exert on one another, the total vector momentum of the system is constant; no violation of this principle has been found. Also known as momentum conservation.

Conservative force field: A field of force in which the work done on a particle in moving it from one point to another depends only on the particle's initial and final positions.

Conservative property: A property with values that do not change in the course of a particular series of events.

Consolute temperature: The upper temperature of immiscibility for a two-component liquid system. Also known as upper consolute temperature; upper critical solution temperature.

Constant-force spring: A spring which has a constant restoring force, regardless of displacement.

Constant of motion: Constant of motion is a quantity

that is conserved throughout the motion,

Constant pressure combustion: Combustion occurring without a pressure change. the compressed air then runs through a combustion chamber, where fuel is burned, heating that air at constant-pressure process.

Constant-speed drive: A mechanism transmitting motion from one shaft to another that does not allow the velocity ratio of the shafts to be varied, or allows it to be varied only in steps.

Constant-velocity universal joint: A universal joint that transmits constant angular velocity from the driving to the driven shaft, such as the Bendix-Weiss universal joint.

Constrained mechanism: A mechanism in which all members move only in prescribed paths.

Construction equipment: Heavy power machines which perform specific construction or demolition functions.

Contact condenser: Contact condensers employ liquid coolants, usually water, which come in direct contact with condensing va-

pors.

Contact-initiated discharge machining: An electromachining process in which the discharge is initiated by allowing the tool and workpiece to come into contact, after which the tool is withdrawn and an arc forms.

Continuous-type furnace: A furnace used for heat treatment of materials, with or without direct firing; pieces are loaded through one door, progress continuously through the furnace, and are discharged from another door.

Continuous brake: A brake which is attached to each car a train, and can be caused to operate in all the cars simultaneously from a point on any car or on the engine

Continuous bucket excavator: A bucket excavator with a continuous bucket elevator mounted in front of the bowl.

Continuous flow conveyor: A totally enclosed, continuous-belt conveyor pulled transversely through a mass of granular, powdered or small-lump material fed from an overhead hopper.

Continuous mixer: Continuous Mixing is the process of continuously metering ingredients directly into the mixing chamber and as a result,

Contour machining: In case of contour machining the work piece surface can be generated in two modes: by "gearing" between tool and work piece or by "sliding" along the work.

Contour turning: Making a three dimensional reproduction of the shape of a template by controlling the cutting tool with a follower that moves over the surface of a template

Contraction: The action or process of becoming smaller or pressed together, as a gas on cooling. the cutting tool axially follows the path with a predefined geometry. Multiple passes of a contouring tool are necessary to create desired contours in the workpiece.

Contrarotating propellers: A pair of propellers on concentric shafts, turning in opposite directions.

Controllable-pitch propeller: An aircraft or ship propeller in which the pitch of the blades can be changed while the propeller is in motion; five types used for aircraft are two-position, variable-pitch, constant-speed, feathering, and reversible-pitch. Abbreviated CP propeller.

Control system: A system in which one or more outputs are forced to change in a desired manner as time progresses.

Convection cooling: Heat transfer by natural, upward flow of hot air from the device being cooled.

Conveyor: Any materials-handling machine designed to move individual articles such as solids or free-flowing bulk materials over a horizontal, inclined, declined, or vertical path of travel with continuous motion.

Cooling coil: A coiled arrangement of pipe or tubing for the transfer of heat between two fluids. the cooling coils are a component formed by tubes of different materials through which a fluid passes, while these have an external contact with the air or a gas, which allows an exchange of heat.

Cooling correction: A correction that must be employed in calorimetry to allow for heat transfer between a body and its surroundings. Also known as radiation correction.

Cooling curve: A cooling curve is a line graph that represents the change of phase of matter, typically from a gas to a solid or a liquid to a solid.

Cooling degree day: Definition of Cooling Degree Day (CDD): The number of cooling degrees in a day is defined as the difference between the mean temperature.

Cooling fin: The fins around the cylinder head of a reciprocating engine, which improve the normal cooling action. The fins increase the area of the cylinder and thus provide for greater heat transfer.

Cooling load: The total amount of heat energy that must be removed from a system by a cooling mechanism in a unit time, equal to the rate at which heat is generated by people, machinery, and processes, plus the net flow of heat into the system not associated with the cooling machinery.

Cooling method: A method of determining the specific heat of a liquid in which the times taken by the liquid and an equal volume of water in an identical vessel to cool through the same range of temperature are compared.

Cooling power: A parameter devised to measure the air's cooling effect upon a human body; it is determined by the amount of heat required

by a device to maintain the device at a constant temperature (usually 34 C); the entire system should be made to correspond, as closely as possible, to the external heat exchange mechanism of the human body.

Cooling range: Difference in temperature of hot water entering and cold water leaving · amount of heat removed by the cooling tower

Cooling stress: Stress resulting from uneven contraction during cooling of metals and ceramics due to uneven temperature distribution.

Coplanar forces: Forces that act in a single plane; thus the forces are parallel to the plane and their points of application are in the plane.

Core drill: A mechanism designed to rotate and to cause an annular-shaped rock-cutting bit to penetrate rock formations, produce cylindrical cores of the formations penetrated, and lift such cores to the surface, where they may be collected and examined.

Coriolis force: A velocity-dependent pseudoforce in a reference frame which is rotating with respect to an inertial reference frame; it is equal and opposite to the product of the mass of the particle on which the force acts and its Coriolis acceleration.

Corliss valve: An oscillating type of valve gear with a trip mechanism for the admission and exhaust of steam to and from an engine cylinder.

Cornice brake: Process in which the sheet is clamped to the work table with an upper wedge-shaped clamping bar while a rotating part bends plate.

Correction chamber: A closable cavity in a weight on an analytical balance; holds material to adjust weight to nominal value.

Corrective maintenance: A procedure of repairing components or equipment as necessary either by on-site repair or by replacing individual elements in order to keep the system in proper operating condition.

Cotter joint: A cotter joint, also known as a socket and spigot joint, is a method of temporarily joining two co-axial rods. One rod is fitted with a spigot, which fits inside a socket on one end of the other rod.

Coulomb friction: Friction occurring between dry surfaces The Coulomb friction model is used for modeling tangential forces between contact surfaces. For one-dimensional problem

Counterbalanced truck: An industrial truck configured so that all of its load during a normal transporting operation is external to the polygon formed by the points where the wheels contact the surface.

Counterblow hammer: A forging hammer in which the ram and anvil are driven toward each other by compressed air or steam.

Counter current flow: A sensible heat-transfer arrangement in which the two fluids flow in opposite directions.

Countersinking: Drilling operation to form a flaring depression around the rim of a hole. Countersinking is using a drill press machine to create a conical hole in your part that matches the angle and head size of a particular screw.

Coupled engine: A locomotive engine having the driving wheels connected by a rod.

Coupled oscillators: A set of particles subject to elastic restoring forces and also to elastic interactions with each other.

Covering power: The degree to which a coating obscures the underlying material. crane: A hoisting machine with a power-operated inclined or horizontal boom and lifting tackle for moving loads vertically and horizontally.

Crane hoist: A mobile construction machine built principally for lifting loads by means of cables and consisting of an undercarriage on which the unit moves, a cab or house which envelops the main frame and contains the power units and controls, and a movable boom over which the cables run.

Crane truck: A crane with a jiblike boom mounted on a truck. Also known as yard crane.

Crank: A crank is an arm attached at a right angle to a rotating shaft by which circular motion is imparted to or received from the shaft. When combined with a connecting rod, it can be used to convert circular motion into reciprocating motion

Crank angle: Crank angle refers to the position of an engine's crankshaft in relation to the piston as it travels inside of the cylinder wall.

Crank axle: An axle containing a crank. An axle bent at both ends so that it can accommodate a large body with large wheels.

Crankcase: The housing for the crankshaft of an engine, where, in the case of an automobile, oil from hot engine parts is collected and cooled before returning to the engine by a pump.

Crank press: A punch press that applies power to the slide by means of a crank

Crankshaft: The shaft about which a crank rotates

Crank throw: The web or arm of a crank. The displacement of a crankpin from the crankshaft.

Crank web: Crank web definition is - the portion of a crank between the crankpin and the shaft or between adjacent crankpins, also called crank arm.

Crater: A depression in the face of a cutting tool worn down by chip contact

Crawler crane: A self-propelled crane mounted on two endless tracks that revolve around wheels.

Crawler wheel: A wheel that drives a continuous metal belt, as on a crawler tractor.

Creep grinding: Creep feed grinding is characterized by lower workpiece speeds and higher depths of cut resulting in a larger arc length of contact between the grinding wheel and workpiece when compared with reciprocating or pendulum grinding.

Creep limit: The maximum stress a given material can withstand in a given time without exceeding a specified quantity of creep.

Creep rupture strength: The stress which, at a given temperature, will cause a material to rupture in a given time.

Creep strength: The stress which, at a given temperature, will result in a creep rate of . 1% deformation within 100,000 hours.

Crinal: A unit of force equal to 0.1 newton.

Crith: A unit of mass, used for gases, equal to the mass of 1 liter of hydrogen at standard pressure and temperature;

Critical compression ratio: The lowest compression ratio which allows compression ignition of a specific fuel.

Critical isotherm: A curve showing the relationship between the pressure and volume of a gas at its critical temperature.

Critical pressure: The pressure of the liquid vapor critical point.

Critical speed: The angular speed at which a rotating shaft becomes dynamically unstable with large lateral amplitudes, due to resonance with the natural frequencies of lateral vibration of the shaft.

Critical vibration: A vibration that is significant and harmful to a structure.

Cross axle: A shaft operated by levers at its ends. An axle with cranks set at 90 .

Cross box: A boxlike structure for the connection of circulating tubes to the longitudinal drum of a header-type boiler.

Cross drum boiler: A sectional header or box header type of boiler in which the axis of the horizontal drum is perpendicular to the axis of the main bank of tubes.

Crossed belt: A pulley belt arranged so that the sides cross, thereby making the pulleys rotate in opposite directions.

Crosshaul: A device for loading objects onto vehicles, consisting of a chain that is hooked on opposite sides of a vehicle, looped under the object, and connected to a power source and that rolls the object onto the vehicle.

Cross slide: A part of a machine tool that allows the tool carriage to move at right angles to the main direction of travel.

Crown sheet: The structural element which forms the top of a furnace in a fire-tube boiler.

Crusher: A machine for crushing rock and other bulk materials.

Crushing strain: Compression which causes the failure of a material.

Crushing strength: The compressive stress required to cause a solid to fail by fracture; in essence, it is the resistance of the solid to vertical pressure placed upon it.

Cryology: The study of low-temperature (approximately 200 R, or -160 °C) refrigeration.

Cryosorption pump: A high-vacuum pump that employs a sorbent such as activated charcoal or synthetic zeolite cooled by nitrogen or some other refrigerant; used to reduce pressure from atmospheric pressure to a few millitorr.

Cubic: Denoting a unit of volume, volume of a cube whose sides have a length of 1 meter. Abbreviated cu.

Cubic boron nitride: A synthetic material composed of boron and nitrogen (1:1) that is almost as hard as diamond, used as a superabrasive powder and for cutting and grinding applications.

Cubic measure: A unit or set of units to measure volume.

Cumec: A unit of volume flow rate equal to 1 cubic meter per second.

Curie principle: The principle that a macroscopic cause never has more elements of symmetry than the effect it produces; for example, a scalar cause cannot produce a vectorial effect.

Curle scale of temperature: A temperature scale based on the susceptibility of a paramagnetic substance, assuming that it obeys Curie's law; used at temperatures below about 1 kelvin

Curling: A forming process in which the edge of a sheet-metal part is rolled over to produce a hollow tubular rim.

Curling dies: A set of tools that shape the ends of a piece of work into a form with a circular cross section.

Curling machine: A machine with curling dies; used to curl the ends of cans

Current line: In marine operations, a graduated line attached to a current pole, used to measure the speed of a current; as the pole moves away with the current, the speed of the current is determined by the amount of line paid out in a specified time.

Curve resistance: The force opposing the motion of a railway train along a track due to track curvature.

Curvilinear motion: Motion along a curved path.

Cutoff point: The point at which there is a transition from spiral flow in the housing of a centrifugal fan to straight-line flow in the connected duct.

Cutoff tool: A tool used on bar-type lathes to separate the finished piece from the bar stock.

Cutoff valve: A valve used to stop the flow of steam to the cylinder of a steam engine.

Cutoff wheel: A thin wheel impregnated with an abrasive used for severing or cut-ting slots in a material or part.

Cutter bar: The bar that supports the cutting tool in a lathe or other machine.

Cutter head: A rotating head or stock, either shaped and ground to form a cutter, or so devised that bits or blades can be attached to it

Cutter sweep: The section that is cut off or eradicated by the milling cutter or grinding wheel in entering or leaving the flute

Cutting angle: It is the angle between the side surface of the tool and a line normal to the base of the tool. (v) End Cutting Edge Angle: It is the angle between the end cutting edge of the tool and a line perpendicular to its shank.

Cutting down: Removing surface roughness or irregularities from metal by the use of an abrasive.

Cutting drilling: Drilling is a cutting process that uses a drill bit to cut a hole of circular cross-section in solid materials. The drill bit is usually a rotary cutting

Cutting in: An undesirable action occurring during loose-drum spooling in which a layer of wire rope spreads apart and forms grooves in which the next layer travels

Cutting-off machine: A machine for cutting off metal bars and shapes; includes the lathe type using single-point cutoff tools, and several types of saws.

Cutting ratio: The Cutting Ratio is defined as the thickness of metal before cutting to the thickness of metal after cutting.

Cutting speed: Cutting speed (also called surface speed or simply speed) is the speed difference (relative velocity) between the cutting tool and the surface of the work piece it is operating on.

Cutting tool: A cutting tool or cutter is any tool that is used to remove some material from the work piece by means of shear deformation

Cyclic coordinate: A generalized coordinate on which the Lagrangian of a system does not depend explicitly.

Cyclic train: A set of gears, such as an epicyclic gear arrangement, in which one or more of the gear axes rotates around a fixed axis.

Cycloidal pendulum: An alteration of a simple pendulum in which a weight is suspended from a cord which is slung between two pieces of metal shaped in the form of cycloids.

Cyclone separator: A funnel-shaped device for removing particles from air or other fluids by centrifugal means; used to re-move dust from air or other fluids, steam from water, and water from steam, and in certain applications to separate particles into two or more size classes. Also known as cyclone classifier.

Cylinder actuator: A device that converts hydraulic power into useful mechanical work by means of a tight-fitting piston moving in a closed cylinder.

Cylinder head: The cylinder head mounts on the cylinder block or engine housing. Together with the piston and cylinder, it forms a part of the combustion chamber.

Cylinder liner: The cylinder liner is a hollow cylindrical shell which acts as the enclosure in which the combustion takes place.

Cylindrical cam: A cylindrical cam is a 3D cam which drives its follower in a groove cut on the periphery of a cylinder. The follower, which is either cylindrical or conical, may translate or oscillate. The

cam rotates about its longitudinal axis, and transmits a
translation or oscillation displacement to the follower at
the same time.

Cylindrical grinder: The
cylindrical grinder is a type of
grinding machine used to
shape the outside of an object.
The cylindrical grinder can
work on a variety of shapes.

Dado head: A rotary cutter composed of several saw-like blades side by side, used for cutting flat-bottomed grooves in wood

D'Alembert's principle: It is an alternative form of Newton's second law of motion. The principle that the resultant of the external forces and the kinetic reaction acting on a body equals zero.

Dalton's temperature scale: A scale intended for measuring temperature such that the absolute temperature T is given in terms of the temperature on the Dalton scale

Damage tolerance: The capability of a structure to maintain its load carrying capability after exposure to a sudden increase in load.

Damaging stress: The least unit stress for a specified material and use that will cause damage to the member and make it unfit for its expected length of service

Damper loss: The decrease in rate of flow or of pressure of gas across a damper.

Damping capacity: A material's ability in absorbing vibrations.

Dandy roll: A roll in a Fourdrinier paper-making machine; inured to compact the sheet and sometimes to imprint a watermark

Datum: A level, direction or position from which angles, heights, speeds or distances are conveniently measured.

Datum plane: A permanently established horizontal plane, surface, or level to which soundings, ground elevations, water surface elevations, and tidal data are referred

Dead axle: An axle that carries a wheel but does not impel it.

Deadbeat: Coming to rest with no vibration or oscillation, as when the pointer of a meter moves to a new position without overshooting. Also known as dead beat response.

Dead block: A tool placed on the ends of railroad passenger cars to absorb the shock of impacts.

Dead center: The extreme position(top and bottom) of a piston in cylinder. A support for the work on a lathe which does not turn with the work.

Deadman's brake: An emergency tool that automatically is activated to stop a vehicle when the driver removes his or her foot from the pedal.

Deadman's handle: A handle on a machine designed so that the operator must continuously press on it in order to keep the machine running.

Dead sheave: A grooved wheel on a crown block over which the deadline is fixed firmly.

Dead-stroke: Having a recoilless or nearly recoilless stroke.

Deaerator: A device that removes oxygen and other dissolved gases from liquids and pumpable compounds.

Decaliter: A unit of volume, equal to 10 liters, or to 0.01 cubic meter.

Decastere: A unit of volume, equal to 10 cubic meters.

Deceleration: The rate of reduce of speed of a motion.

Deciare: A unit of area, equal to 0.1 are or 10 square meters.

Decibar: A metric unit of pressure equal to one-tenth bar.

Decigram: A unit of mass, equal to 0.1 gram

Decilitre: A metric unit of capacity, equal to one tenth of a litre.

Decimetre: A metric unit of length equal to one-tenth meter

Decompression: Any procedure for the relief of pressure or compression.

Deconcentrator: An equipment for removing dissolved or suspended material from feed-water

Decremeter: A tool for measuring the logarithmic decrement (damping) of a train of waves.

Dedendum: The difference between the pitch circle radius of the of the gear and the radius of its root circle.

Deep-well pump: A multistage centrifugal pump for lifting water from profound, small-diameter wells.

Deflecting torque: An instrument's moment, resulting from the quantity measured, that acts to reason the pointer's deflection

Deflection curve: The curve, normally downward, described by a shot deviating from its true course.

Deflection meter: A deflection meter consists of a vane or plate that projects into the flow and a sensing element that measures the deflection caused by the force of the flow against the vane.

Deformation: Deformation refers to the change in size or shape of an object due to external force.

Deformation curve: A curve showing the relation-ship between the stress and the strain. Also known as stress-strain curve.

Degradation: The conver-sion of energy into forms that are increasingly difficult to convert into work, resulting from the general tendency of entropy to increase.

Degree: One of the units of temperature or temperature difference in any of various temperature scales, such as the Celsius, Fahrenheit, and Kelvin temperature scales.

Degree-day: Degree days are measures of how cold or warm a location is. A degree day compares the mean (the average of the high and low) outdoor temperatures rec-orded for a location to a standard temperature.

Degree of freedom: Nu-meral independent motions that are allowed to the body or, in case of

a mechanism made of sev-eral bodies, number of pos-sible independent relative motions between the pieces of the mechanism.

Dehumidification: The process of reducing the moisture in the air. It serves to in-crease the cooling pow-er of air.

Dehumidifier: Apparatus designed to reduce the amount of water vapor in the ambient atmosphere.

Demister: A series of pipes in automobiles arranged so that hot, dry air directed from the heat source is forced against the interior of the windscreen or windshield to stop condensation.

Demon of Maxwell: Demon of Maxwell is a thought experiment that would hypothetically violate the second law of thermody-namics. Also known as Max-well's demon.

Densify: The method to in-crease the density of a materi-al.

Densimeter: A device which measures the density or spe-cific gravity of a liquid, gas, or solid.

Density: The mass of a given substance per unit volume (Kg/m^3).

Depth gage: A device or tool for measuring the depth of depression to a thousandth inch.

Descending branch: That portion of a trajectory which is between the summit and the point where the trajectory terminates, either by impact or air burst, and along which the projectile falls, with altitude constantly decreasing.

Deslimer: It is an Apparatus, like a bowl-type centrifuge, used to remove fine, wet particles (slime) from cement rocks and to size pigments and abrasives

Detent: A catch or lever in a mechanism which initiates or locks motion of a part, especially in escapement mechanisms

Detonating rate: The velocity at which the explosion wave go by through a cylindrical charge.

Detonation: Impulsive combustion of the compressed charge after passage of the spark in an internal combustion engine; it is accompanied by knock.

Detonation front: The reaction zone of a detonation.

Deviation: The variation of the actual value of a controlled variable and the desired value corresponding to the set point.

Deviatoric stress: Deviatoric stress is the difference between the stress tensor σ and hydrostatic pressure tensor p acting on the rock or soil mass.

Dewatering: Mechanical dewatering is normally associated with large wastewater treatment plants and is used to separate sludge (residual sludge from wastewater treatment plants or faecal sludge from on-site sanitation) .

Diabatic: When a system changes from one state to another by transfer of heat across the boundaries of the system. Also known as non-adiabatic.

Diagonal stay: A boiler stay is an internal structural element used inside boilers. A diagonal member between the tube sheet and shell in a fire-tube boiler.

Diagram factor: The diagram factor is the ratio of the area of actual indicator diagram to the area of theoretical indicator diagram.

Dial feed: An apparatus that rotates work pieces into position successively so they can be acted on by a machine

Dial press: A punch press with dial feed.

Diameter group: A dimensionless group, used in the study of flow machines such as turbines and pumps.

Diaphragm compressor: Tool for compression of small volumes of a gas by means of a reciprocally moving diaphragm, in place of pistons or rotors

Diaphragm pump: A metering pump which uses a diaphragm to isolate the operating parts from pumped liquid in a mechanically actuated diaphragm pump, or from hydraulic fluid in a hydraulically actuated diaphragm pump.

Die block: A tool-steel block which is bolted to the bed of a punch press and into which the desired impressions are machined.

Die body: The stationary element of an extrusion die, used to separate and form material.

Dieing machine: A vertical press with the slide activated by pull rods attached to the drive mechanism below the bed of the press.

Diesel cycle: An internal combustion engine cycle in which heat addition takes place at constant pressure.

Diesel electric locomotive: A locomotive with a diesel engine driving an electric generator which provides electric power to traction motors for propelling the vehicle. Also known as diesel locomotive.

Diesel electric power generation: Electric power generation in which the generator is driven by a diesel engine.

Diesel engine: An internal combustion engine operating on a thermodynamic cycle in which the ratio of compression of the air charge is sufficiently high to ignite the fuel subsequently injected into the combustion chamber. Also known as compression-ignition engine.

Dieseling: Explosions of mixtures of air and lubricating oil in the compression chambers or in other component of the air system of a compressor.

Diesel knock: A combustion knock caused when the delayed period of ignition is long so that a large quantity of atomized fuel accumulates in the combustion chamber; when combustion occurs, the sudden high pressure resulting from the accumulated fuel causes diesel knock.

Diesel rig: Any diesel engine apparatus or machinery.

Die shoe: A metal block inserted between the lower half of a cutting or shaping die and the bed of a press to spread the blow and avoid wear.

Die slide: An apparatus in which the lower die of a power press is mounted; it slides in and out of the press for easy access and safety in feeding the parts

Differential brake: A brake in which the action depends on a difference between two motions

Differential effects: The effects upon the elements of the trajectory due to variations from standard conditions.

Differential indexing: A method of subdividing a circle based on the difference among movements of the index plate and index crank of a dividing engine

Differential motion: A mechanism in which the follower has two driving elements; the net motion of the follower is the difference between the motions that would result from either driver acting alone.

Differential pulley: A differential pulley, also called "Weston differential pulley", sometimes "chain hoist" or colloquially "chain fall", is used to manually lift very heavy objects like car engines. It is operated by pulling upon the slack section of a continuous chain that wraps around pulleys. The relative size of the two connected pulleys determines the maximum weight that can be lifted by hand.

Differential screw: A type of compound screw which create a motion equal to the difference in motion between the two component screws.

Differential windlass: A windlass in which the barrel has two sections, each having a different diameter; the rope winds around one section, passes through a pulley (which carries the load), then winds around the other section of the barrel.

Diffusivity: Diffusivity is a rate of diffusion. The quantity of heat passing normally through a unit area per unit time divided by the product of specific heat, density, and temperature gradient. It is also known as thermal diffusivity; thermometric conductivity.

Dings magnetic separator: A device which is suspended above a belt conveyor to pull out and separate magnetic material from burden as thick as 40 inches (1 meter) and at belt speeds up to 750 feet (229 meters) per minute.

Dinking: Using a sharp, hollow punch for cutting lightgage soft metals or nonmetallic materials.

Dipper dredge: A power shovel resembling a grab crane mounted on a flat bottom boat for dredging under water. Also known as dipper shovel.

Dipper stick: A straight shaft linking the digging bucket of an excavating machine or power shovel with the boom.

Dipper trip: A tool which releases the door of a shovel bucket.

Direct acting pump: A displacement reciprocating pump where the steam or power piston is connected to the pump piston by means of a rod, without crank motion or flywheel.

Direct connected: The link between a driver and a driven part, as a turbine and an electric generator, without intervening speed changing devices, such as gears.

Direct-coupled: connected without intermediate connections.

Direct drive: A drive in which the driving part is directly connected to the driven part.

Direct drive vibration machine: A vibration appliance in which the vibration table is forced to undergo a displacement by a positive linkage driven by a direct attachment to eccentrics or camshafts.

Direct-expansion coil: A finned coil, used in air cooling, inside of which circulates a cold fluid or evaporating refrigerant.

Direct-geared: Joined by a gear on the shaft of one machine meshing with a gear on the shaft of another machine.

Directional gyro: A two-degrees-of-freedom gyro with a provision for maintaining its spin axis approximately horizontal.

Direct return system: In a heating or cooling arrangement, a piping arrangement in which the fluid is returned to its origin (boiler or evaporator) by the shortest direct path after it has passed through each heat exchanger.

Discharge channel: The

way in a pressure relief device through which the fluid is released to the outside of the device.

Discharge head: Vertical distance between the intake level of a water pump and the level at which it discharges water freely to the atmosphere.

Discharge line: The length of a duct through which drilling mud travels from the mud pump through the standpipe on its way to the borehole.

Disintegrator: An apparatus used for pulverizing or grinding substances, consisting of two steel cages which rotate in opposite directions.

Disk brake: A kind of brake in which disks attached to a fixed frame are pressed against disks attached to a rotating axle or against the inner surfaces of a rotating housing.

Disk cam: A disk with a contoured edge which rotates about an axis perpendicular to the disk, communicating motion to the cam follower which remains in contact with the edge of the disk.

Disk centrifuge: A centrifuge with a big bowl having a set of disks that split the liq-

uid into thin layers to create shallow settling chambers.

Disk clutch: A clutch in which torque is transmitted by friction between friction disks with specially prepared friction material riveted to both sides and contact plates keyed to the inner surface of an external hub.

Disk coupling: A flexible coupling in which the connecting member is a flexible disk.

Disk grinder: A grinding apparatus that employs abrasive disks.

Disk grinding: Grinding with the flat side of a rigid, bonded abrasive disk or segmental wheel.

Disk mill: Size reduction apparatus in which grinding of feed solids takes place between two disks, either or both of which rotate.

Disk sander: A machine that employs a circular disk coated with abrasive to smooth or shape surfaces.

Disk spring: A mechanical spring that consists of a disk or washer supported by one force (distributed by a suitable chuck or holder) at the periphery and by an opposing force on the center or hub of the disk.

Dispersion mill: Size-reduction apparatus that disrupts clusters or agglomerates of solids, rather than breaking down individual particles, used for paint pigments, food products, and cosmetics.

Displacement compressor: A type of compressor that depends on displacement of a volume of air by a piston moving in a cylinder.

Displacement pump: A pump that build up its action through the alternate filling and emptying of an covered volume as in a piston cylinder construction.

Distance: The spatial parting of two points, measured by the length of a hypothetical line joining them.

Distance ratio: The ratio of the distance moved by the effort or input of a machine in a specified time to the distance moved by the load or output.

Disk mill: Size reduction equipment in which grinding of feed solids takes place between two disks, either or both of which rotate.

Disk spring: A mechanical spring that consists of a disk or washer supported by one force (distributed by a suitable chuck or holder) at the periphery and by an opposing force   on the center or hub of the disk.

Dispersion mill: Size-reduction apparatus that disrupts clusters or agglomerates of solids, rather than breaking down individual particles; used for paint pigments, food products, and cosmetics

Distributor gear: A gear which meshes with the camshaft gear to turn the distributor shaft.

District heating: The supply of heat, either in the form of steam or hot water, from a central source to a group of buildings

Divariant system: A system composed of only one phase, so that two variables, such as pressure and temperature, are sufficient to define its thermodynamic state.

Division plate: A diaphragm which surrounds the piston rod of a crosshead type engine and separates the crankcase from the lower part of the cylinder.

Domestic refrigerator: A refrigeration system for household use which typically has a compression machine designed for continuous automatic operation and for conservation of the charges of refrigerant and oil, and is usually motor driven and air cooled.

Donkey engine: A small supplementary engine which is usually moveable and powered by steam.

Donohue equation: Equation used to determine the heat transfer film coefficient for a fluid on the outside of a baffled shell-and-tube heat exchanger.

Doodlebug: A small tractor. A motor-driven railcar used for maintenance and repair work.

Dorr classifier: A horizontal flow classifier consisting of a rectangular tank with a sloping bottom, a rake mechanism for moving sands uphill along the bottom, an inlet for feed, and outlets for sand and slime.

Double acting: A double-acting cylinder is a cylinder in which the working fluid acts alternately on both sides of the piston.

Double acting compressor: A reciprocating compressor in which both sides of the piston perform in working chambers to compress the fluid.

Double acting pawl: A dou-ble pawl which can drive in either direction.

Double action mechanical press: A press having two slides which move one within the other in parallel movements.

Double block brake: Two single block brakes in symmetrical opposition, where the operating force on one lever is the reaction on the other.

Double crank press: A mechanical press with a single wide slide operated by a crankshaft having two crank pins

Double-cut planer: A planer designed to cut in both the forward and reverse strokes of the table.

Double-drum hoist: A hoisting device consisting of two cable drums which rotate in opposite directions and can be operated separately or together.

Double Hooke's joint: A universal joint which eradicate the variation in angular displacement and angular velocity among driving and driven shafts.

Double-housing planer: A planer having two housings to support the cross rail, with two heads on the cross rail and one side head on each housing.

Double integrating gyro: A single degree of freedom gyro having essentially no restraint of its spin axis about the output axis.

Double roll crusher: An apparatus which crushes materials between teeth on two roll surfaces, used mainly for coal.

Double weighing: A method of weighing to allow for differences in lengths of the balance arms, in which object and weights are balanced twice, the second time with their positions inter changed. Also known as Gauss method of weighing.

Downdraft carburettor: A carburetor in which the fuel is supply into a downward current of air.

Down feed system: In a heating or cooling arrangement, a piping arrangement in which the fluid is circulated through supply mains that are located above the levels of the units they serve.

Draft loss: A decrease in the static pressure of a gas in a furnace or boiler due to flow resistance.

Draft tube: The piping arrangement for a reaction-type hydraulic turbine that allows the turbine to be set safely above tail water and yet utilize the full head of the site from head race to tail race.

Drag chain conveyor: A conveyor in which the open links of a chain drag material along the bottom of a hard faced concrete or cast iron trough. Also known as dragline conveyor.

Drag classifier: A continuous belt containing transverse rakes, used to separate coarse sand from fine; the belt moves up through an inclined trough, and fast-settling sands are dragged along by the rakes.

Dragline: An excavator controlled by pulling a bucket on ropes towards the jib from which it is suspended. Also known as dragline excavator

Dragline scraper: An apparatus with a flat, plow like blade or partially open bucket pulled on rope for withdrawing piled material, such as stone or coal, from a stockyard to the loading platform; the empty bucket is subsequently returned to the pile of material by means of a return rope.

Drag link: A drag link converts rotary motion from a crank arm, to a second bell crank, usually in an automotive steering system.

Dram: A unit of weight in the avoirdupois system equal to one sixteenth of an ounce.

Drawbar pull: The force by which a locomotive or tractor pulls vehicles on a draw bar behind it.

Dressler kiln: The first successful muffle type tunnel kiln.

Drifter: A rock drill, similar to but usually larger than a jack hammer, mounted for drilling holes up to 41/2 inches (11.4 centimetres) in diameter.

Drill capacity: The length of drill rod of particular size that the hoist on a diamond or rotary drill can lift or that the brake can hold on a single line.

Drill feed: The method by which the drill bit is fed into the borehole during drilling.

Drilling machine: An apparatus, usually motor driven, fitted with an end cutting tool that is rotated with sufficient power either to create a hole or to enlarge an existing hole in a solid material. Also known as driller.

Drilling rate: The speed at which a drill bit breaks the rock under it to deepen the borehole. It is normally measured in feet per minute or meters per hour.

Drill jig: A tool fastened to the work in repetition drilling to position and guide the drill

Drill press: A drilling machine in which a vertical drill moves into the work, which is stationary.

Drill string: The assemblage of drill rods, core barrel, and bit, or of drill rods, drill collars, and bit in a borehole, which is connected to and rotated by the drill collar of the borehole.

Drive chuck: A method at the lower end of a diamond drill drive rod on the swivel head by means of which the motion of the drive rod can be transmitted to the drill string.

Driveline: In an automotive vehicle, the group of parts, including the universal joint and the drive shaft, that connect the transmission with the driving wheels.

Driving pinion: The input gear in the differential of an automobile.

Driving resistance: The force exerted by soil on a pile being driven into it.

Driving wheel: A wheel that provides driving power.

Droop governor: A governor whose equilibrium speed reduces as the load on the machinery controlled by the governor amplifies.

Dropwise condensation: Condensation of a vapor on a face in which the condensate forms into drops.

Drum brake: A brake in which two curved shoes fitted with heat and wear resistant linings are forced against the surface of a rotating drum.

Drum dryer: A machine for removing water from substances such as milk, in which a thin film of the product is moved over a turning steam-heated drum.

Drum feeder: A rotating drum with vanes or buckets to raise and take parts and drop them into various orienting or chute arrangements. Also known as tumbler feeder.

Drum filter: A cylindrical drum that turns through thickened ore pulp, extracts liquid by a vacuum, and leaves solids, in the form of a cake, on a permeable membrane on the drum end.

Dry abrasive cutting: Frictional cutting using a rotary abrasive wheel without the use of a liquid coolant.

Dry cooling tower: An arrangement in which water is cooled by circulation through finned tubes, transferring heat to air passing over the fins; there is no loss of water by evaporation because the air does not directly contact the water.

Dry friction: Resistance between two dry solid surfaces, that is, surfaces free from contaminating films or fluids.

Dry measure: A measure of volume for commodities that are dry.

Dry mill: Grinding device used to powder or pulverize solid materials without an associated liquid.

Dry pipe: A perforated metal pipe above the normal water level in the steam space of a boiler which prevents moisture or extraneous matter from entering steam outlet lines

Dry pit pump: A pump operated with the liquid conducted to and from the unit by piping.

Dry steam drum: Pressurized chamber into which steam flows from the steam space of a boiler drum.

Dry storage: Cold storage in

which refrigeration is provided by chilled air.

Dry strength: The strength of an adhesive joint determined immediately after drying under specified conditions or after a period of conditioning in the standard laboratory atmosphere

Dual bed dehumidifier: A sorbent dehumidifier with two beds, one bed dehumidifying while the other bed is reactivating, thus providing a continuous flow of air.

Dual flow oil burner: An oil burner with two sets of tangential slots in its atomizer for use at different capacity levels.

Dual fuel engine: Internal combustion engine that can function on either of two fuels, such as natural gas or gasoline.

Duckbill: A shaking type of combination loader and conveyor whose loading end is generally shaped like a duck's bill.

Duct: Ducts are conduits or passages used in heating, ventilation, and air conditioning to deliver and remove air.

Ducted fan: A propeller or multi bladed fan inside a coaxial duct or cowling. Also known as ducted propeller.

Dumbwaiter: An industrial elevator which transmit small objects but is not permitted to carry people.

Dump bucket: A big bucket with movable discharge gates at the bottom, used to move soil or other construction materials by a crane or cable.

Dump car: Any of several types of narrow gage rail cars with bodies which can easily be tipped to dump material.

Duplex pump: A reciprocating pump with two parallel pumping cylinders.

Duplex tandem compressor: A compressor having cylinders on two parallel frames connected through a common crankshaft.

Durability: The quality of equipment, structures, or goods of continuing to be useful after an extended period of time and usage.

Duration: A basic concept of kinetics which is expressed quantitatively by time measured by a clock or comparable method.

Dutchman: A filler piece for closing a gap among two pipes or between a pipe or fitting and a piece of equipment, if the pipe is too short to achieve closure or if the pipe and equipment are not aligned.

Dutch process: A process for making white lead, metallic lead is placed in vessels containing a dilute acetic acid, and the vessels are stacked in bark or manure.

Dynamical similarity: Two flow fields are dynamically similar if one can be transformed into the other by a change of length and velocity scales. All dimensionless numbers of the flows must be the same.

Dynamical variable: A quantity used to explain a system in classical mechanics, such as the coordinates of a particle, the components of its velocity, and momentum, or the functions of these quantities.

Dynamic augment: Force created by unbalanced reciprocating parts in a steam locomotive.

Dynamic balance: The condition that exists in a rotating body when the axis it is forced to rotate around, or which is being pointed, is parallel to a major axis of inertia; There are no products of inertia about the center of gravity of the body with respect to the specified axis of rotation.

Dynamic behaviour: A description of how an individual system or unit works with respect to time.

Dynamic braking: A technique of electric braking in which the retarding force is supplied by the same machine that originally was the driving motor.

Dynamic compressor: A compressor which employ rotating vanes or impellers to impart velocity and pressure to the fluid.

Dynamic creep: Creep resulting from fluctuations in a load or temperature.

Dynamic equilibrium: The state of any mechanical system when the kinetic interaction is considered as a force, such that the resultant force on the system is zero according to D'Alembert's principle. Also known as kinetic balance.

Dynamics: Branch of mechanics that deals with the motion of a system of material particles under the influence of forces, especially those that arise outside the system under study.

Dynamic similarity: A relationship between two mechanical systems (often referred to as a model and a prototype) such that by relative adjustments of the units of length, mass, and time, the quantities measured in one system are identical (or to a constant multiple of each) in those in the other; In particular, this means that there are constant ratios of forces in the two systems.

Dynamic stability: The characteristic of a body, such as an aircraft, rocket, or ship, that causes it, when disturbed from an original state of steady motion in an upright position, to damp the oscillations set up by restoring moments and gradually return to its original state. Also known as stability.

Dynamic unbalance: Failure of the axis of rotation of a piece of rotating equipment to match one of the principal axis of inertia due to forces in one axial plane and on opposite sides of the axis of rotation, or in different axial planes.

Dyne: The unit of force in the system of units is in grams per second, and the force that gives an acceleration of 1 cm/sec is equal to 2 to a mass of 1 gram.

Earliest finish time: The earliest time to complete project work; For the entire project, it equals the earliest start time for the final event listed in the table.

Earliest start time: The earliest an action in the project plan can begin; It equals the earliest time that all previous activities can be completed.

Early finish date: The earliest time that an activity can be completed.

Earthmover: A device used for excavating, transporting, or pushing the central eccentricity of the Earth. The distance of the geometric center of the rotating body from the axis of rotation.

Eccentric load: A load imposed on a structural portion at some point other than the centroid of the section.

Eccentric reducer: A threaded or butt welded fitting for pipes whose ends are not the same size and are eccentric to each other.

Eccentric rotor engine: A rotary motor, such as a Wankel motor, in which motion is transmitted to the shaft by an eccentric rotor to the shaft.

Eccentric valve: A style of rotary control valve with a plug-shaped flow-restricting member that follows an eccentric path as it rotates. The plug does not contact its seat until it rotates within a few degrees of the closed position.

Economizer: A mechanical devices intended to reduce energy consumption, or to perform useful function such as preheating a fluid.

Eddy conductivity: A control device or dynamometer for regulating rotational speed, as of flywheels, in which energy is converted by eddy currents into heat.

Eddy-current clutch: A type of electromagnetic clutch in which torque is conveyed by means of eddy currents induced by a magnetic field set up by a coil carrying direct current in one rotating member.

Eddy-current sensor: A proximity sensor which uses an alternating magnetic field to create eddy currents in nearby objects, and then the currents are used to detect the presence of the objects.

Eddy heat conduction: Heat transfer by vortices in turbulent flow, treated similarly to molecular conduction. Also known as eddy heat flow; Eddy delivery.

Effective launcher line: The line that would run parallel to the aircraft missile if it was not induced by gravity.

Effective rake: The angular relationship among the plane of the tooth face of the cutter and the line through the tooth point measured in the direction of chip stream.

Effective surface: In a heat exchanger, a plane that keenly transmit heat.

Ejector condenser: A kind of direct contact condenser in which the vacuum is maintained by high-speed injection water; Steam condenses and discharges water, condensate and non-condensate into the atmosphere.

Elastic: Capable of sustaining deformation without unending loss of size or shape.

Elastic after effect: When an elastic body is stretched and applied deforming force is removed then the body is expected to return to its original configuration immediately.

Elastic axis: The lengthways line of a beam along which transverse loads must be applied in order to produce bending only, with no torsion of the beam at any part.

Elastic body: The body that regain its original shape and size on the removal of deforming.

Elastic buckling: A sudden raise in the lateral deflection of a column at a critical load while the stresses acting on the column are wholly elastic.

Elastic center: The elastic center is the point at which an applied force produces pure translation, and an applied moment produces pure rotation about the same axis.

Elastic collision: A collision in which the sum of the kinetic energies of translation of the participating systems is the same after the collision as before.

Elastic curve: The curved shape of the longitudinal centroidal surface of a beam when the transverse loads acting on it produced wholly elastic stresses.

Elastic deformation: Elastic deformation refers to a temporary deformation of a material's shape that is self-reversing after removing the force or load. Elastic deformation alters the shape of a material upon the application of a force within its elastic limit.

Elastic equilibrium: The state of an elastic body in which each volume component of the body is in equilibrium under the combined influence of elastic stresses and externally applied body forces.

Elastic failure: Failure of a body to recover its original size and shape after a stress is removed.

Elastic flow: Return of a material to its original shape following deformation.

Elastic force: A force arising from the deformation of a solid body which depends only on the body's instantaneous deformation and not on its previous history, and which is conservative.

Elastic hysteresis: A phenomenon shown by some solids in which the deformation of the solid depends not only on the stress applied to the solid, but also on the previous history of that stress; Similar to magnetic hysteresis, the magnetic field strength and magnetic induction are replaced by stress and strain respectively.

Elasticity: The property whereby a solid material changes its shape and size under action of opposing forces, but recovers its original con-

figuration when the forces are removed.

Elastic potential energy: Capacity that a body has to do work by virtue of its deformation.

Elastic ratio: The ratio of the limit of elasticity to the final strength of a solid.

Elastic recovery: A specific deformation of a solid behaves in an elastic. Scattering due to elastic collision.

Elastic strain energy: A theory of relationships between forces acting on an object and resulting changes in dimensions.

Elastic vibration: The oscillatory motion of a rigid body supported by the elastic and inertial forces of the body.

Elasto dynamics: The study of the mechanical properties of elastic waves.

Elasto plasticity: State of a matter subjected to a stress greater than its elastic limit but not so great as to cause it to rupture, in which it exhibits both elastic and plastic properties.

Electrical weighing system: An apparatus which weighs an object by measuring the change in resistance caused by the elastic deformation of a mechanical element loaded with the object.

Electric boiler: A steam generator with electric energy, in immersion, resistor, or electrode elements, as the source of heat.

Electric brake: An actuator in which the actuating force is supplied by current flowing through a solenoid.

Electric car: An automotive vehicle that is propelled by one or more electric motors powered by a special rechargeable electric battery rather than by an internal combustion engine.

Electric cell: A single unit of a primary or secondary battery that transfer chemical energy into electric energy.

Electric circuit : Also known as circuit. A path or group of interconnected paths able to carrying electric currents

Electric coupling: Magnetic field coupling among the shafts of a driver and a driven machine.

Electric drive: A method which transmits motion from one shaft to another and controls the velocity ratio of the shafts by electrical means.

Electric field: One of the fundamental fields in nature, causing a charged body to be attracted to or repelled by other charged bodies; associated with an electromagnetic wave or a changing magnetic field.

Electric furnace: A furnace which employ electricity as a source of heat.

Electric hammer: An electric powered hammer, often used for riveting or caulking.

Electric heating: A process of converting electric energy to heat energy by resisting the free flow of electric current.

Electric ignition: Ignition of a charge of fuel vapor and air in an internal combustion engine

Electric locomotive: A locomotive operated by electric power picked up from a system of continuous overhead wires, or, some-times, from a third rail mounted alongside the track.

Electric power generation: The large scale production of electric power for industrial, residential, and rural use.

Electric power plant: A power plant that generate electricity from a source. The sources are like water, steam, diesel, or nuclear.

Electric power system: A complex grouping of tools and circuits for generating, transmitting, transforming, and distributing electric energy.

Electric railroad: A railroad which has a system of continuous overhead wires or a third rail mounted alongside the track to supply electric power to the locomotive and cars.

Electric stacker: A stacker whose carriage is raised and lowered by a winch powered by electric storage batteries.

Electric thermometer: An apparatus that utilizes electrical means to measure temperature, such as a thermocouple or resistance thermometer.

Electric vehicle: A ground vehicle propelled by a motor powered by electrical energy from rechargeable batteries or other source onboard the vehicle, or from an external source in, on, or above the roadway; examples include the electrically powered golf cart, automobile, and trolley bus.

Electrochemical thermodynamics: The application of the laws of thermodynamics to electrochemical systems.

Electro drill: A drilling machine driven by electric power.

Electrolytic grinding: A combined grinding and machining process in which the abrasive, cathodic grinding wheel is in contact with the anodic work piece.

Electro machining: The appliance of electric or ultrasonic energy to a work piece to effect removal of material.

Electromagnetic clutch: A clutch based on magnetic coupling between conductors, such as a magnetic fluid and powder clutch, Example: eddy-current clutch, or a hysteresis clutch.

Electromechanical: Pertaining to a mechanical device, system, or process which is electro statically or electro magnetically actuated or controlled.

Electro mechanics: The technology of mechanical devices, systems, or processes which are electro statically or electromagnetically actuated or controlled.

Electrostriction: A form of elastic deformation of a dielectric induced by an electric field.

Elements: The various characteristics of a trajectory such as the angle of departure, maximum ordinate, angle of fall, and so on.

Elevation: Vertical distance to a point or object from sea level or some other datum.

Elevation meter: An apparatus that measures the change of elevation of a vehicle

Elliptical orbit: The path of a body moving along an ellipse, such as that described by either of two bodies revolving under their mutual gravitational attraction but otherwise undisturbed.

Elliptic gear: A gear shift consisting of two elliptical gears, each of which rotates around its axis points.

Emagram: An emagram is one of four thermodynamic diagrams used to display temperature lapse rate and moisture content profiles in the atmosphere. The emagram has axes of temperature and pressure.

Embrittlement: Reduction or loss of ductility or toughness in a metal or plastic with little change in other mechanical properties.

Emergency brake: A brake that can be set by hand and, once set, continues to hold until released; used as a parking brake in an automobile.

Emissivity : The ratio of the radiation emitted by a surface to the radiation emitted by a perfect blackbody radiator at the same temperature. Also known as thermal emissivity.

Emittance: The power radiated per unit area of a radiating surface. Also known as emissive power; radiating power

End-feed centerless grinding: Centerless grinding in which the piece is fed through grinding and regulating wheels to an end stop.

End loader: A platform elevator at the rear of a truck.

End mill : A machine which has a rotating shank with cutting teeth at the end and spiral blades on the peripheral surface; used for shaping and cutting metal

End play: Axial movement in a shaft and bearing assembly resulting from clearances between the components.

End stop: A limit to the movement of a mechanical system or part, usually brought about by valves or shock absorbers.

Energy beam: An intense beam of light, electrons, or other nuclear particles; used to cut, drill, form, weld, or otherwise process metals, ceramics, and other materials.

Energy conversion efficiency: The efficiency with which the energy of the working material is converted into kinetic energy.

Energy integral: In the case of a conservative force, the integral constant obtained by integrating Newton's second law of motion; equal to the sum of the kinetic energy of the particle and the potential energy of the force acting on it.

Engine: A machine in which power is applied to do work by the conversion of various forms of energy into mechanical force and motion.

Engine balance: Arrangement and construction of moving element in reciprocating or rotating machines to decrease dynamic forces which may effect in undesirable vibrations

Engine cooling: Controlling the temperature of internal combustion engine element to avoid overheating and to maintain all operating dimensions, clearances, and alignment by a circulating coolant, oil, and a fan.

Engine cycle: Any series of thermodynamic phases constituting a cycle for the conversion of heat into work, examples are the Otto cycle, Stirling cycle, Dual cycle and Diesel cycle.

Engine cylinder: A cylindrical hollow space in an engine in which the energy of the working fluid, in the form of pressure and heat, is converted to mechanical force by performing work on the piston. Also known as cylinder.

Engine displacement: Volume displaced by each piston moving from bottom dead center to top dead center multiplied by the number of cylinders.

Engine efficiency: Ratio between the energy supplied to an engine to the energy output of the engine.

Engine inlet: A position of entrance for engine fuel.

Engine knock: In spark ignition engines, the sound and other property associated with ignition and rapid combustion of the last part of the charge to burn, before the flame front reaches it. Also known as combustion knock.

Engine lathe: A manually

operated lathe equipped with a headstock of the back geared, cone-driven type or of the geared-head type.

Engine performance: Relationship between power output, revolutions per minute, fuel or fluid consumption, and ambient conditions in which an engine work.

Entering angle: The angle between the side cutting edge of a tool and the machined surface of the work; angle is 90 for a tool with 0 side cutting edge angle effective.

Enthalpy: The sum of the internal energy of a system plus the product of the system's volume multiplied by the pressure exerted on the system by its surroundings. Also known as heat content; sensible heat; total heat.

Entropy: Entropy is a scientific concept, as well as a measurable physical property that is most commonly associated with a state of disorder, randomness, or uncertainty

Entry ballistics: That branch of ballistics which pertains to the entry of a missile, space craft, or other object from outer space into and through an atmosphere.

Environment: The aggregate of all natural, operational, or other conditions that affect the operation of tools or components.

Environmental cab: Operator's section in earthmovers equipped with tinted safety glass, sound proofing, air conditioning, and cleaning units.

Environmental control: changes and control of soil, water, and air environments of humans and other living organisms

Environmental engineering: The technology concerned with the reduction of pollution, contamination, and deterioration of the surroundings in which humans live.

Environmental protection: The protection of humans and equipment against stresses of climate and other elements of the environment

Epicyclic gear: A planetary or **epicyclic gear** train. A system of gears in which one or more gears travel around the inside or the outside of another gear whose axis is fixed.

Epicyclic train: A combination of epicyclic gears, usually connected by an arm, in which some or all of the gears have a motion compounded of rotation about an axis and a translation or revolution of that axis.

Equal-arm balance: A simple balance in which the distance from the support point of the balance arm beam to the two pans at the end of the beam is equal.

Equation of piezotropy: An equation obeyed by certain fluids which states that the time rate of change of the fluid's density equals the product of a function of the thermodynamic variables and the time rate of change of the pressure.

Equatorial plane: A plane perpendicular to the axis of rotation of a rotating body and equidistant from the intersections of this axis with the body's surface, provided that the body is symmetric about the axis of rotation and is symmetric under reflection through this plane.

Equilibrant: A single force which revoke the vector sum of a given system of forces acting on a rigid body and whose torque cancels the sum of the torques of the system.

Equilibrium: Circumstance in which a particle, or all the constituent particles of a body, are at rest or in un accelerated motion in an inertial reference frame. Also known as static equilibrium.

Equipment: One or more assemblage capable of performing a complete function.

Equivalent bending moment: A bending moment which, acting alone, would produce in a circular shaft a normal stress of the same magnitude as the maximum normal stress produced by a given bending moment and a given twisting moment acting simultaneously.

Equivalent blackbody temperature: For a surface, the temperature of a blackbody which emits the same amount of radiation per unit area as does the surface.

Equivalent nitrogen pressure: The pressure that would be specify by a device if the gas inside it were replaced by nitrogen of equivalent molecular density

Equivalent orifice: An expression of fan performance as the theoretical sharp edge orifice area which would offer the same resistance to flow as the system resistance itself.

Equivalent temperature: A

term used in British engineering for that temperature of a uniform enclosure in which, in still air, a sizable blackbody at 75 F (23.9 C) would lose heat at the same rate as in the environment.

Equivalent twisting moment: A twisting moment which, if acting alone, would produce in a circular shaft a shear stress of the same magnitude as the shear stress produced by a given twisting moment and a given bending moment acting simultaneously.

Equivalent viscous damping: The hypothetical value used to analyze the viscous damping of the vibration motion, so that the energy dissipation of each cycle at resonance is the same for the hypothetical or actual damping force.

Erection stress: The internal forces exerted on a structural member during construction.

Ericsson cycle: An ideal thermodynamic cycle consisting of two isobaric processes interspersed with processes which are, in effect, isothermal, but each of which consists of an infinite number of alternating isentropic and isobaric processes.

Escalator: A continuously moving stairway and handrail.

Escapement: A ratchet device that allows motion in one direction slowly.

Euler angles: Three angular parameters that identify the orientation of a body with respect to reference axes.

Euler equation: Euler's equation for a steady flow of an ideal fluid along a streamline is **a relation between the velocity, pressure, and density of a moving fluid**. It is based on Newton's Second Law of Motion.

Euler equations of motion: A set of three differential equations expressing relations between the force moments, angular velocities, and angular accelerations of a rotating rigid body.

Euler force: The greatest load that a long, slender column can carry without buckling, according to the Euler formula for long columns

Euler formula for long columns: A formula which gives the greatest axial load that a long, slender column can carry without buckling, in terms of its length, Young's modulus, and the moment of inertia about an axis along the center of the column.

Euler method: A process of studying fluid motion and the mechanics of deformable bodies in which one considers volume elements at fixed locations in space, across which material flows; the Euler method is in contrast to the Lagrangian method.

Euler-Rodrigues parameter: One of four numbers which may be used to specify the orientation of a rigid body; they are components of a quaternion.

Evaporative condenser: An apparatus in which vapor is condensed within tubes that are cooled by the evaporation of water flowing over the outside of the tubes

Evaporative control system: A motor vehicle system that prevents escape of gasoline vapors from the fuel tank or carburetor to the atmosphere while the engine is not operating

Evaporative cooling: Lowering the temperature of a large mass of liquid by utilizing the latent heat of vaporization of a portion of the liquid.

Excavator: A machine for digging and removing earth

Exergy: The portion of the total energy of a system that is available for conversion to useful work

Exhaust: The working substance discharged from an engine cylinder or turbine after performing work on the moving parts of the machine.

Exhaust deflecting ring: A jet aircraft consisting of a ring installed at the end of the nozzle to allow it to spin into the exhaust stream.

Exhaust gas: Spent gas leaving an internal combustion engine or gas turbine.

Exhaust manifold: A branched system of pipes to carry waste emissions away from the piston chambers of an internal combustion engine.

Exhaust pipe: The duct through which engine exhaust is discharged

Exhaust stroke: The stroke of an engine, pump, or compressor that expels the fluid from the cylinder.

Exhaust suction stroke: A stroke of an engine that simultaneously removes used fuel and introduces fresh fuel to the cylinder

Exhaust valve: The valve on a cylinder in an internal combustion engine which controls the discharge of spent gas.

Expanding brake: A brake that operates by moving outward against the inside rim of a drum or wheel.

Expansion cooling: Cooling of a substance by having it undergo adiabatic expansion.

Expansion engine: Piston-cylinder device that cools compressed air via sudden expansion; used in production of pure gaseous oxygen via the Claude cycle.

Expansion ratio: In a reciprocating piston engine, the ratio of cylinder volume with piston at bottom dead center to cylinder volume with piston at top dead center.

Expansion valve: A valve in which fluid flows under falling pressure and increasing volume.

Explosion door: A door in a furnace which is designed to open at a predetermined pressure.

Explosion method: A method of measuring the specific heat of a constant volume gas. This method is to wrap an explosive mixture with a known heat of reaction in the gas, place it in a chamber enclosed by a corrugated steel membrane as a pressure gauge, and then deduce that the mixture is ig-

nited due to pressure changes The highest temperature reached at the time.

Explosion rupture disk device: A protective tool used where the pressure rise in the vessel occurs at a rapid rate.

Extensibility: The amount to which a material can be stretched or distorted without breaking

Exterior ballistics: The science concerned with behavior of a projectile after leaving the muzzle of the firing weapon.

External brake: A brake that operates by contacting the outside of a brake drum

External centerless grinding: A process by which a metal workpiece is finished on its external surface by supporting the piece on a blade while it is advanced between a regulating wheel and grinding wheel.

External combustion engine: An engine in which the generation of heat is effected in a furnace or reactor outside the engine cylinder.

External force: A force exerted on a system or on some of its components by an agency outside the system

External grinding: Grinding the outer surface of a rotating piece of work

External header: Manifold connecting sections of a cast iron boiler

Externally fired boiler: A boiler that has refractory or cooling tubes surrounding its furnace

External shoe brake: A friction brake operated by the application of externally contracting elements

External work: The work done by a system in expanding against forces exerted from outside

Extruder: A device that forces ductile or semisoft solids through die openings of appropriate shape to produce a continuous film, strip, or tubing

Extrusion: A process in which a hot or cold semisoft solid material, such as metal or plastic, is forced through the orifice of a die to produce a continuously formed piece in the shape of the desired product.

Extrusion coating: A process of placing resin on a substrate by extruding a thin film of molten resin and pressing it onto or into the substrates, or both, without

the use of adhesives

Fabrication: Metal fabrication is the creation of metal structures by cutting, bending and assembling processes. It is a value-added process involving the creation of machines, parts, and structures from various raw materials.

Face discharge bit: A liquid coolant bit intended for drilling in soft formations and for use on a double-tube core barrel

Face milling: Milling flat surfaces perpendicular to the axis of rotation of the cutting tool.

Face shield: A removable wraparound guard fitted to a worker's helmet to protect the face from flying particles.

Factor of safety: The ratio of the ultimate strength of a member or piece of material to the actual working stress or the maximum permissible stress when in use.

Factor of stress concentration: Any irregularity producing localized stress in a structural member subject to load.

Factor of stress intensity: The ratio of the maximum stress to which a structural member can be subjected to the maximum stress to which

it is likely to be subjected.

Fahrenheit scale: A temperature scale, the temperature in degrees Fahrenheit (F)

Failed hole: A drill hole loaded with dynamite which did not blow up. Also known as missed hole.

Failure rate: The probability of failure per unit of time of items in operation.

Fairlead: A group of pulleys or rollers used in conjunction with a winch or similar apparatus to allow the cable to be reeled from any direction.

Fales Stuart windmill: A windmill developed for farm use from the two blade airfoil propeller.

Falk flexible coupling: A spring coupling in which a continuous steel spring is threaded back and forth through axial slots in the periphery of two hubs on the shaft ends.

Fall block: A pulley block that rises and falls with the load on a lifting tackle.

Faller: A machine part whose operation depends on a falling action.

Falling body: A body whose motion is accelerated towards the center of the

Earth under the influence of gravity; other forces acting on it are negligible in comparison with it.

Fan: A device, usually consisting of a rotating impeller or propeller, with or without a shroud, to generate currents to circulate, release, or deliver large volumes of air or gas.

Fan brake: A fan used to provide a load for a driving mechanism.

Fan efficiency: The ratio obtained by dividing a fan's useful power output by the power input.

Fan rating: The head, quantity, power, and efficiency expected from a fan operating at peak efficiency.

Fan shaft: The spindle on which a fan impeller is mounted.

Fan static pressure: The total pressure rise diminished by the velocity pressure in the fan outlet.

Fan test: Observations of the quantity, total pressure, and power of air circulated by a fan running at a known constant speed.

Fan total head: The sum of the fan static head and the velocity head at the fan discharge corresponding to a giv-en quantity of air flow.

Fan velocity pressure: The velocity pressure corresponding to the average velocity at the fan outlet.

Fast coupling: A flexible gear coupling that uses two inner hubs on circumferential gear shafts, surrounded by an internally-toothed gear shroud to engage and connect the two hubs.

Fatigue life: The number of applied repeated stress cycles a material can endure before failure.

Fatigue limit: The maximum stress that a material can endure for an infinite number of stress cycles without breaking. Also known as endurance limit.

Fatigue ratio: The ratio of the fatigue limit or fatigue strength to the static tensile strength.

Fatigue strength: The maximum stress a material can endure for a given number of stress cycles without breaking. Also known as endurance strength

Feather: To change the pitch on a propeller in order to reduce drag and prevent wind milling in case of engine failure.

Feathering: A pitch position in a controllable pitch propeller, it is used in the event of engine failure to stop the wind milling action

Feed control valve: A small valve, usually a needle valve, on the outlet of the hydraulic-feed cylinder on the swivel head of a diamond drill, used to control minutely the speed of the hydraulic piston travel and hence the rate at which the bit is made to penetrate the rock.

Feeder-breaker: A unit that breaks and feeds ore or crushed rock to a materials-handling system at a required rate.

Feeder conveyor : A short auxiliary conveyor designed to transport materials to another conveyor. Also known as stage loader

Feed nut: The threaded sleeve fitting around the feed screw on a gear feed drill swivel head, which is rotated by means of paired gears driven from the spindle or feed shaft.

Feed pipe: The pipe which conducts water to a boiler drum.

Feed pump: A pump used to supply water to a steam boiler.

Feed ratio: The number of revolutions a drill stem and bit must turn to advance the drill bit 1 inch.

Feed screw: The externally threaded drill rod drive rod in a screw- or gear-feed swivel head on a diamond drill; also used on percussion drills, lathes, and other machinery.

Feed shaft: A stub shaft or countershaft in a gear-driven diamond drill feed pivot head that is rotated by an electric motor on a drill rig through gears or a partial drive and which drives a meshed pair of feed gears.

Feed travel: Distance that the drilling machine moves the steel shank from top to bottom in the feed range.

Feed trough: A receptacle into which feed water overflows from a boiler drum.

Feed water: The water supplied to a boiler or still.

Feed water heater: An apparatus that utilizes steam extracted from an engine.

Feeler gage: A tool with many blades of different thickness used to establish clearance between parts or for gapping spark plugs.

Feeler pin: A pin that allows a duplicating machine to operate only when there is a supply of paper.

Femtometer: A unit of length, equal to 10^{-15} meter; used particularly in measuring nuclear distances. Abbreviated fm. Also known as fermi.

Fiber stress: The tensile or compressive stress on the fibers of a fiber metal or other fibrous material, especially when fiber orientation is parallel with the neutral axis.

Fibrous fracture: Failure of a material resulting from a ductile crack; broken surfaces are dull and silky.

Field excitation: Controlling the speed of a series motor in an electric or diesel electric locomotive by changing the ratio between armature current and field strength, either by reducing the excitation current by shunting the excitation coils with resistances, or by using field taps.

Fifth wheel: A coupling device in the form of two horizontal disks that revolve on each other positioned between a tractor and a semi-trailer so that they can change direction independently.

Fill factor: The estimated load that the dipper of a shov-

el is carrying, expressed as a percentage of the rated capacity.

Film boiling: Boiling in which a constant film of vapor forms at the hot surface of the container holding the boiling liquid, reducing heat transfer across the surface.

Film coefficient: For a fluid confined in a vessel, the rate of flow of heat out of the fluid, per unit area of vessel wall divided by the difference between the temperature in the interior of the fluid and the temperature at the sur-face of the wall.

Film condensation: The formation of a continuous film of liquid on a wall in contact with a vapor, when the wall is cooled below the local vapor saturation temperature and the liquid wets the cold surface.

Film cooling: The cooling of a body or surface, such as the inner surface of a rocket combustion chamber by maintaining a thin fluid layer over the affected area.

Film transport: The method for moving photographic film through the region where light strikes it in recording film tracks or sound tracks of motion pictures.

Filter pump: An aspirator or vacuum pump which generate a negative pressure on the filtrate side of the filter to hasten the process of filtering.

Fine grinding: Grinding performed in a mill rotating on a horizontal axis in which the material undergoes final size reduction.

Finish grinding: The last action of a grinding process to achieve a good finish and accurate dimensions.

Finish turning: The operation of treating a surface to a precise size and obtaining a smooth surface.

Finite strain theory: A theory of elasticity, suitable for high compressions, in which it is not assumed that strains are infinitesimally small.

Finned surface: A tubular heat exchange surface with complete projections on one side.

Fire pump: A pump for fire protection purposes usually driven by an independent, reliable prime mover and approved by the National Board of Fire Underwriters.

Fire room: That portion of a fossil fuel-burning plant which contains the furnace and associated equipment.

Fire-tube boiler: A steam boiler in which hot gaseous products of combustion pass through tubes surrounded by boiler water.

Firing machine: An electric blasting machine.

Firing pressure: The maximum pressure in an engine cylinder during combustion.

Firing rate: The rate at which fuel feed to a burner occurs, in terms of volume, heat units, or weight per unit time.

Firmoviscosity: Property of a matter in which the stress is equal to the sum of a term proportional to the substance's deformation, and a term proportional to its rate of deformation.

First law of thermodynamics: The law states that, heat is a form of energy, and the total amount of energy of all kinds in an isolated system is constant; it is an application of the principle of conservation of energy.

First order transition: A change in state of aggregation of a system accompanied by a discontinuous change in enthalpy, entropy, and volume at a single temperature and pressure.

Fishing: In drilling, the operation by which lost or damaged tools are secured and brought to the surface from the bottom of a well or drill hole.

Fishing tool: A device for retrieving objects from inaccessible locations.

Fixed cost: A cost that remains unchanged during short term changes in production level. Also known as overhead; overhead cost.

Fixed end: An end of a structure, such as a beam, that is clamped in place so that both its position and orientation are fixed.

Fixed feed grinding: Feeding processed material to a grinding wheel, or vice versa, in predetermined increments or at a given rate.

Fixing moment: The bending moment at the end support of the beam is necessary to fix it and prevent rotation. Also known as fixed end point.

Flaking mill: A machine for converting material to flakes.

Flame detector: A sensing device that indicates whether the fuel is on or if the ignition is lost by transmitting a signal to the control system.

Flanging: A forming process in which the edge of a metal part is bent over to make a flange at a sharp angle to the body of the part.

Flap valve: A valve fitted with a hinged flap or disk that swings in one direction only.

Flash: In plastics or rubber molding or in metal casting, that portion of the charge which over flows from the mold cavity at the joint line.

Flash boiler: Small hot tube boiler; designed to immediately convert a small amount of water into superheated steam.

Flash line: A raised line on the surface of a molding where the mold faces joined.

Flat belt conveyor: Conveyor belt in which the carrier section is supported by tension rollers or flat belt pulleys.

Flat blade turbine: An impeller with flat blades attached to the margin.

Flat spin: The movement of the projectile with slow rotation and a very large yaw angle, which most often occurs in projectiles with fin stabilization with some torque, when the period of rotation of the projectile coincides with the period of its oscillations; sometimes seen in bombs and unstable rotating projectiles.

Flat trajectory: A relatively flat trajectory, that is, described by a projectile at a relatively high speed.

Flat-turret lathe: A lathe with a low, flat turret on a power-fed cross-sliding head stock.

Fleet: Sidewise movement of a rope or cable when winding on a drum.

Fleet angle: In the hoisting mechanism - the angle between the rope in the position of the greatest stroke along the drum and a line drawn perpendicular to the drum shaft, passing through the center of the head pulley or the groove of the driving pulley.

Flettner windmill: An inefficient windmill with four arms, each consisting of a rotating cylinder actuated by a Savonius rotor.

Flexibility: The quality or state of being able to be flexed or bent repeatedly.

Flexible coupling : A coupling used to attach two shafts and to accommodate their misalignment.

Flexible shaft: A shaft that transmits rotary motion at any angle up to about 90 .

Flexural rigidity: The ratio of the sideward force applied to one end of a beam to the resulting displacement of this end, when the other end is clamped.

Flexural strength: Strength of a material in blending, that is, resistance to fracture.

Flexure: The deformation of any beam subjected to a load.

Flexure theory: The theory of deformation of a prismatic beam having a length at least 10 times its depth and consisting of a material obeying Hooke's law in response to stresses within elastic limits.

Flight conveyor: Conveyor in which blades, attached to single or double strands of chain, pull or push shredded or granular solids along a chute. Also known as scraper conveyor.

Flight feeder: Short length flap conveyor used to feed solids into a process vessel or other container at a predetermined speed.

Flint mill: A mill employing pebbles to pulverize materials.

Float bowl: A carburetor component that contains a small amount of liquid gasoline and serves as a constant level reservoir of fuel that is metered into the passing air stream.

Float gage: Any of several types of instruments in which the level of a liquid is determined by the height of a body floating on its surface using rods, levers, or other mechanical devices.

Floating axle: Driving axle used to turn the wheels of a motor vehicle; The weight of the vehicle is borne by the covers at the ends of the fixed axle.

Floating lever: A horizontal brake lever with a movable fulcrum; used under railroad cars.

Floating scraper: A balanced scraper blade that rests lightly on a drum filter; removes solids collected on the rotating drum surface by riding on the drum's surface contour.

Float level: The position of the float in a carburetor at which the needle valve closes the fuel inlet to prevent entry of additional fuel

Flotation: A process used to separate particulate solids by causing one group of particles to float

Flow: A forward movement in a continuous stream or sequence of fluids or discrete objects or materials, as in a continuous chemical process or solids-conveying or production-line operations.

Flow chart symbol: Any of the existing symbols commonly used to represent operations, data, or material flow, or equipment in a data processing task or manufacturing process description.

Flow control: Any system used to control the flow of gases, vapors, liquids, slurries, pastes, or solid particles through or along conduits or channels.

Flow control valve: A valve whose flow opening is controlled by the rate of flow of the fluid through it; usually controlled by differential pressure across an orifice at the valve. Also known as rate-of-flow control valve.

Flow curve: The stress-strain curve of a plastic material.

Flowing-temperature factor: Calculated correction factor for gases flowing at temperatures other than that for which the flow equation holds, that is, at temperatures other than 60 F (15.5 C).

Flow line: The connecting line or arrow between symbols on a flow chart or block diagram.

Flow measurement: Determining the amount of a fluid, be it liquid, vapor, or gas, that passes through a pipe, conduit, or open conduit.

Flow meter: An apparatus used to measure pressure, flow rate, and discharge rate of a liquid, vapor, or gas flowing in a pipe. Also known as fluid meter.

Flow mixer: A device for mixing liquids and liquids in which mixing occurs as liquids pass through it; includes jet nozzles and agitator blades.

Flow process: A system in which liquids or solids are in continuous motion during chemical or physical processing or manufacturing.

Flow-rating pressure: The value of inlet static pressure at which the relieving capacity of a pressure relief device is established.

Flow stress: The stress along one axis at a given value of strain that is essential to produce plastic deformation.

Flow visualization: Method of making visible the disturbances that occur in fluid flow, using the fact that light passing through a flow field of varying density exhibits refraction and a relative phase shift among different rays.

Flue: A channel or passage for assigning combustion products from a furnace, boiler, or fireplace to or through a chimney.

Flue gas expander: In an oil refining system, a turbine for energy recovery at the point where combustion gases are released under pressure into the atmosphere; the decrease in pressure drives the turbine impeller.

Fluid controlled valve: A valve for which the valve operator is activated by a fluid energy, in contrast to electrical, pneumatic, or manual energy.

Fluid coupling: A tool for transmitting rotation between shafts by means of the acceleration and deceleration of a fluid such as oil. Also known as hydraulic coupling.

Fluid die: A die for shaping parts by liquid pressure; a plunger forces the liquid against the part to be shaped, making the part conform to the shape of a die.

Fluid dram: A unit of volume used in the United States for measurement of liquid substances.

Fluid drive: A power coupling operate on a hydraulic turbine principle in which the engine flywheel has a set of turbine blades which are connected directly to it and which are driven in oil, thereby turning another set of blades attached to the transmission gears of the automobile

Fluid end: In a fluid pump, the section that surround parts which are directly involved in moving the fluid.

Fluid-film bearing: A rolling bearing in which the friction surfaces are separated by a film of lubricant, such as oil.

Fluidized bed combustion: A method of burning particulate fuels such as coal, in which the amount of air required for combustion is much higher than the amount of air in conventional burners.

Fluid mechanics: The science anxious with fluids, either at rest or in motion, and dealing with pressures, velocities, and accelerations in the fluid, including fluid deformation and compression or expansion.

Fluid ounce: A unit of volume that is used in the United States for measurement of liquid substances.

Fluid stress: Stress allied with plastic deformation in a solid material.

Fluid ton: A unit of volume equal to 32 cubic feet or approximately 0.90614 cubic meter; used for many hydrometallurgical, hydraulic, and other industrial purposes.

Fluid transmission: Automotive transmission with fluid drive.

Fluting: A machining method whereby flutes are formed parallel to the main axis of cylindrical or conical parts.

Flutter: The irregular alternating movement of the parts of a relief valve due to the application of pressure where no contact is made between the valve disk and the seat.

Fly cutter: A cutting tool that revolves with the arbor of a lathe.

Fly cutting: Cutting with a milling cutter provided with only one tooth.

Flywheel: A flywheel is a mechanical device which uses the conservation of angular momentum to store rotational energy; a form of kinetic energy proportional to the product of its moment of inertia and the square of its rotational speed.

Foaming: Any of the various processes by which air or gas is introduced into a liquid or solid to form a foam.

Foot: The unit of length in the British systems of units, equal to exactly 0.3048 meter.

Footage: The extent or length of a material expressed in feet.

Foot-pound: A unit of energy or work in the English gravitational system, equal to the work done by 1 pound of force when the point of application of the force is displaced 1 foot in the direction of the force.

Foot-poundal: A unit of energy or work in the English absolute system, equal to the work done by a force of magnitude 1 poundal when the point at which the force is applied is displaced 1 foot in the direction of the force; equal to approximately 0.04214011 joule

Foot section: In both belt and chain conveyors part of the conveyor at the extreme opposite end from the delivery point.

Footstock: An apparatus containing a center which supports the work piece on a milling machine; usually used in conjunction with a dividing head.

Foot valve: A valve in the bottom of the suction pipe of a pump which avoid backward flow of water.

Forbes bar: A metal rod, one end of which is immersed in a molten metal crucible, and thermometers placed in holes at regular intervals along the rod; measuring the temperature along the rod together with the measurement of the cooling of a short section of the rod allows you to calculate the thermal conductivity of the metal.

Force: That influence on a body which causes it to accelerate; quantitatively it is a vector, equal to the body's time rate of change of momentum

Force constant: The ratio of force to deformation of a system whose deformation is proportional to the applied force.

Forced air heating: A warm air heating method in which positive air circulation is provided by means of a fan or a blower.

Forced circulation: Using a pump or other fluid transfer device in conjunction with fluid handling equipment to move fluid through pipes and process vessels;

as opposed to gravity or heat circulation

Forced circulation boiler: A once through steam generator in which water is pumped through successive parts.

Forced convection: Forced convection is a special type of heat transfer in which fluids are forced to move, in order to increase the heat transfer. This forcing can be done with a ceiling fan, a pump, suction device, or other.

Forced draft: Air under positive pressure formed by fans at the point where air or gases enter a unit, such as a combustion furnace.

Forced oscillation: An oscillation formed in a simple oscillator or equivalent mechanical system by an external periodic driving force.

Forced ventilation: A system of ventilation in which air is forced through ventilation ducts under pressure.

Force polygon: A closed polygon whose sides are vectors representing the forces acting on a body in equilibrium.

Force pump: A pump equipped with a solid plunger and a suction valve that draws in and pumps liquid to

a significant height above the valve or creates significant pressure on the liquid.

Forklift: A machine, usually powered by hydraulic means, consisting of two or more tines that can be raised and lowered and inserted under heavy materials or objects to lift and move them.

Fork pocket: An opening in the base of a container or pallet for insertion of the prong of a forklift.

Fork truck: A vehicle equipped with a forklift. Also known as forklift truck.

Formed cutter: A cutting tool designed to create surfaces with irregular geometry. Also known as a milling machine.

Form grinding: Grinding using a wheel whose cutting surface is the opposite of the desired shape.

Forming die: A die like a drawing die, but without a blank holder.

Forming press: A punch press for forming metal parts.

Forming rolls: Rolls contoured to give a desired shape to parts passing through them.

Foucault pendulum: A

swinging load supported by a long rope so that the top support of the rope only holds the rope in the vertical direction and the load is swinging without lateral or circular movement.

Foundry: A building where metal or glass castings are produced.

Foundry engineering: The science and practice of melting and casting glass or metal.

Four-bar linkage: A plane linkage consisting of four links pinned tail to head in a closed loop with lower, or closed, joints

Fourdrinier machine: Paper machine; the paper web is formed on an endless wire mesh; the screen passes through presses and dryers to calenders and drums.

Fourier law of heat conduction: The law according to which the rate of heat flow through a substance is proportional to the area perpendicular to the direction of the flow and the negative rate of temperature change with distance along the direction of the flow. Also known as the Fourier heat equation.

Fourier number: Dimensionless number used in the study of unsteady heat transfer, equal to the product of thermal conductivity and characteristic time divided by the product of density, specific heat at constant pressure and distance from the midpoint. a body through which heat passes to the surface.

Four-stroke cycle: An internal combustion engine cycle completed in four piston strokes; includes suction stroke, compression stroke, expansion stroke and discharge stroke.

Four-way valve: A valve at the junction of four waterways which allows passage between any two adjacent waterways by means of a movable element operated by a quarter turn.

Four-wheel drive: An arrangement in which the drive shaft acts on all four wheels of the automobile.

Fox lathe: A lathe with chasing bar and leaders for cutting threads; used for turning brass.

Fracture stress: The minimum tensile stress that will cause fracture. Also known as fracture strength.

Fracture test: Macro or microscopic examination of a

fractured surface to determine characteristics such as grain pattern, composition, or the presence of defects

Fracture wear: The wear on individual abrasive grains on the surface of a grinding wheel caused by fracture.

Framework: The load carrying frame of a structure; may be of timber, steel, or concrete

Francis turbine: A reaction hydraulic turbine of relatively medium speed with radial flow of water in the runner.

Free energy: The internal energy of a system minus the product of its temperature and its entropy. Also known as Helmholtz free energy

Free fall: The ideal falling motion of a body acted upon only by the pull of the earth's gravitational field.

Free falling: In ball milling, the peripheral speed at which part of the crop load breaks clear on the ascending side and falls clear to the toe of the charge.

Free flight: Unconstrained or unassisted flight.

Free-flight angle: The angle between the horizontal and a line in the direction of motion of a flying body, especially a

rocket, at the beginning of free flight.

Free-flight trajectory: The path of a body in free fall.

Free joint: A robotic articulation that has six degrees of freedom

Free-piston engine: A prime mover utilizing free-piston motion controlled by gas pressure in the cylinders.

Free turbine: In a turbine engine, a turbine wheel that drives the output shaft and is not connected to the shaft driving the compressor.

Free vector: A vector whose direction in space is prescribed but whose point of application and line of application are not prescribed.

Freezer: An insulated unit, compartment, or room in which perishable foods are quick-frozen and stored.

Freeze-up: Abnormal operation of the refrigeration unit due to the formation of ice on the expansion device.

Frequency meter: An instrument for measuring the frequency of an alternating current.

Friction: A force that opposes the relative motion of two bodies when such a motion exists or when there are other forces that tend to produce such a motion.

Frictional grip: Coupling the wheels of the locomotive with the rails of the railway track.

Friction bearing: A solid bearing that directly contacts and supports an axle end.

Friction brake: A brake in which the resistance is provided by friction.

Friction clutch: A clutch in which torque is transmitted by pressure of the clutch faces on each other.

Friction drive: A drive that work by the friction forces set up when one rotating

Friction gear: Gearing in which motion is transmitted through friction between two surfaces in rolling contact.

Friction horse power: Power dissipated in a machine through friction.

Friction loss: Loss of mechanical energy due to mechanical friction between moving parts of the machine.

Friction saw: A toothless circular saw used to cut materials by fusion due to frictional heat.

Friction sawing: Firing process for cutting the workpiece to the desired length using a hacksaw at high speed; especially used for mild steel and stainless steel structural parts.

Friction shoe: An adjustable friction device that holds the sash in any desired state. open position.

Friction torque: Torque generated by frictional forces that counteracts rotary motion, such as those associated with plain or plain bearings in machines.

Friction welding: A process for welding metals and thermoplastics in which two elements are joined by rubbing mating surfaces against each other under high pressure.

Frigorie: A unit of rate of extraction of heat used in refrigeration, equal to 1000 fifteen-degree calories per hour, or 1.16264 0.00014 watts.

Front end loader: Excavator consisting of an articulated bucket mounted on several movable arms at the front of a track or tractor with rubber tires.

Frosting: Decorating a scraped metal surface with a hand scraper. Also known as flaking.

Fuel bed: A layer of burning fuel, as on a furnace grate or a cupola.

Fuel filter: A device in an internal combustion engine, that removes particles from the fuel.

Fuel injection: The release of fuel to an internal combustion engine cylinder by pressure from a mechanical pump.

Fuel injector: A pump method that sprays fuel into the cylinder of an internal combustion engine at the appropriate part of the cycle.

Fuel pump: A pump for drawing fuel from a storage tank and delivering it to an engine or furnace

Fuel system: An arrangement which stores fuel for present use and delivers it as needed.

Fuel tank: The operating, fuel-storage component of a fuel system.

Fugacity: A function used as an analog of the partial pressure in applying thermodynamics to real systems.

Fugacity coefficient: The ratio of the fugacity of a gas to its pressure.

Fulcrum: The rigid point of a support about which the lever pivots.

Full gear: The state of the steam engine when the valve is maximally actuated by the movement of the link.

Full-track vehicle: The vehicle is fully supported, propelled and driven by an endless belt or track on each side.

Full trailer: A towed vehicle whose weight rests completely on its own wheels

Fundamental interval: A value arbitrarily assigned to the temperature difference between two fixed points (such as ice point and vapor point) on the temperature scale to define the scale.

Funicular polygon: A figure formed by a light rope suspended between two points, to which weights are suspended at different points.

Furlong: A unit of length, equal to 1/8 mile, 660 feet, or 201.168 meters.

Fusibility: The quality or degree of liquefaction when heated.

Fusing disk: A fast rotating disc that melts the metal.

G: A unit of acceleration equal to the standard acceleration of gravity, 9.80665 meters per second square.

Gage pressure: The quantity by which the total absolute pressure exceeds the ambient atmospheric pressure.

Gal: The unit of acceleration in the centimeter-gram-second system, equal to 1 centimeter per second squared; commonly used in geodetic measurement.

Galitzin pendulum: A massive horizontal pendulum that is used to calculate variations in the direction of the force of gravity with time, and thus serves as the basis of a seismograph.

Gang drill: A set of drills operated together in the same machine; used in rock drilling.

Gang milling: Rolling of material by means of a composite machine with numerous cutting blades.

Gang saw: A steel frame in which thin, parallel saws are in order to operate concurrently in cutting logs.

Gantry crane: A bridge like hoisting machine having fixed supports or arranged for running along tracks on ground level.

Gap-frame press: A punch press whose frame is open at bed level so that wide work or strip work can be inserted.

Gap lathe: An engine lathe with a sliding bed providing enough space for turning large diameter work.

Gas bearing: A journal or thrust bearing lubricated by gas. Also known as gas lubricated bearing.

Gas burner: A opening or a group of openings through which a combustible gas or gas-air mixture flows and burns.

Gas compression cycle: A refrigeration cycle in which hot compressed gas is cooled in a heat exchanger then passes into a gas expander which provides an exhaust stream of cold gas to another heat exchanger that handles the sensible heat refrigeration effect and exhausts the gas to the compressor.

Gas compressor: A machine that increases the pressure of a gas or vapor by increasing the gas density and delivering the fluid against the associated system resistance.

Gas constant: The constant

of proportionality appearing in the equation of state of an ideal gas, equal to the pressure of the gas times its molar volume divided by its temperature.

Gas cycle: A series in which a gaseous fluid undergoes a series of thermodynamic phases, ultimately returning to its original state.

Gas cylinder: The cavity in which a piston moves in a positive displacement engine or compressor.

Gas engine: An internal combustion engine that uses gaseous fuel.

Gas heater: A unit heater intended to supply heat by forced convection, using gas as a heat source.

Gas injection: Inoculation of gaseous fuel into the cylinder of an internal combustion engine at the appropriate part of the cycle.

Gas law: Any law relating the pressure, volume, and temperature of a gas.

Gas manometer: A gage for determining the difference in pressure of two gases, usually by measuring the difference in height of liquid columns in the two sides of a U-tube.

Gasoline engine: An internal combustion engine that utilize a mixture of air and gasoline vapor as a fuel.

Gasoline pump: A tool that pumps and measures the gasoline supplied to a motor vehicle, as at a filling station.

Gas tank: A tank for storing gas or gasoline.

Gas thermometry: Measurement of temperatures with a gas thermometer, used with helium down to about 1 K.

Gas turbine: A heat engine that changes the energy of fuel into work by using compressed, hot gas as the working medium and that usually delivers its mechanical output power either as torque through a rotating shaft.

Gas turbine nozzle: The part of a gas turbine in which the hot, high pressure gas expands and accelerates to high velocity.

Gas valve: An exhaust valve, held shut by rubber springs, used to release gas from the extreme top of a balloon.

Gates crusher: A gyratory crusher having a cone or mantle that is moved eccentrically by the lower bearing sleeve.

Gate valve: A valve with a disk shaped closing element

that fits tightly over an opening through which water overtakes.

Gear case: An inclusion usually filled with lubricating fluid, in which gears operate.

Gear cutter: A machine or tool for cutting teeth in a gear.

Gear cutting: The cutting or forming of a uniform series of tooth like projections on the surface of a work piece.

Gear down: To organize gears so the driven part rotates at a slower speed than the driving part.

Gear drive: Transmission of motion or torque from one shaft to another by means of direct contact among toothed wheels.

Geared turbine: A turbine connected to a set of reduction gears.

Gear forming: A method of gear cutting in which the desired tooth shape is shaped by a tool whose cutting profile matches the tooth form.

Gear generating: A method of gear cutting in which the tooth is produced by the conjugate or total cutting action of the tool plus the rotation of the work piece.

Gear grinding: A gear cutting method in which

gears are shaped by formed grinding wheels and by generation, primarily a finishing operation.

Gear hobber: A machine that mills gear teeth, the revolving speed of the hob has a precise relationship to that of the work.

Gearing: A set of gear wheels, used to transmit motion from one toothed wheel or sprocket to another.

Gear level: To organize gears so that the driven part and driving part turn at the same speed.

Gear loading: The power transmitted or the contact force per unit length of a gear.

Gear motor: A motor united with a set of speed reducing gears.

Gear pump: A rotary pump in which two meshing gear wheels contra rotate so that the fluid is entrained on one side and discharged on the other.

Gear ratio: The ratio of the angular velocity of the driving member of a gear train or similar mechanism to the angular velocity of the driven member; in particular, the number of engine revolutions per revolution of the rear wheels of the vehicle.

Gear shaper: A machine that build gear teeth by means of a reciprocating cutter that rotates slowly with the work.

Gear shaving machine: finishing machine that take away excess metal from machined gears by the axial sliding motion of a straight-rack cutter or a circular gear cutter.

Gearshift: A tool for engaging and disengaging gears.

Gear train: A blend of two or more gears used to transmit motion between two rotating shafts or between a shaft and a slide.

Gear up: To organize gears so that the driven part rotates faster than the driving part.

Gear wheel: A wheel that meshes gear teeth with another part.

Generalized coordinates: A set of variables used to specify the position and orientation of an arrangement.

Generalized force: The generalized force corresponding to a generalized coordinate is the ratio of the virtual work done in an infinitesimal virtual displacement, which alters that coordinate and no other, to the change in the coordinate.

Generalized velocity: Time derivative of one of the generalized coordinates of the particle. Also known as generalized Lagrangian velocity.

Generating station: A stationary plant containing apparatus for large-scale conversion of some form of energy (such as hydraulic, steam, chemical, or nuclear energy) into electrical energy.

Geographical mile: The length of 1 minute of arc of the Equator, or 6087.08 feet (1855.34 meters), which approximates the length of the nautical mile.

Giaque's temperature scale: The internationally accepted scale of absolute temperature, in which the triple point of water is defined to have a temperature of 273.16 K.

Gib: A detachable plate designed to hold other parts in place or act as a bearing or wear surface.

Gill: A unit of volume used in the United States for the measurement of liquid substances

Gin: A hoisting machine in the form of a tripod with a windlass, pulleys, and ropes.

Gin pole: A hand operated derrick which has a nearly vertical pole supported by guy ropes, the load is raised on a rope that passes through a pulley at the top and over a winch at the foot.

Gin tackle: A tackle made for use with a gin.

Glazed: Pertaining to an abrasive surface that has become smooth and cannot abrade efficiently.

Globe valve: An instrument for regulating flow in a pipeline, consisting of a movable disk type element and a stationary ring seat in a generally spherical body.

Glow plug: A small electric heater, located in a cylinder of a diesel engine, that preheats the air and aids the engine in starting.

Glue joint ripsaw: A heavy gage ripsaw used on straight line or self-feed rip machines, the cut is smooth enough to permit gluing of joints from the saw.

Glug: A unit of mass, equal to the mass which is accelerated by 1 centimetre per second per second by a force of 1 gram-force, or to 980.665 grams.

Goggles: Spectacle like eye protectors having shields at the sides and short, projecting eye tubes.

Gold point: The temperature of the freezing point of gold at a pressure of 1 standard atmosphere (101.325 pascals), used to define the International Temperature Scale of 1940, on which it is assigned a value of 1337.33 K or 1064.18 C.

Governor: A device driven by the centrifugal force of rotating weights opposing gravity or springs, used to provide automatic control of the speed or power of the prime mover.

Grabbing crane: An excavator consisting of a crane carrying a large grapple or bucket in the form of a pair of half-scoop halves hinged to dig into the ground while lifting.

Grab bucket: A bucket with articulated jaws or teeth that is suspended from the cables of a crane or excavator and is used for digging and lifting materials.

Grab dredger: Dredging equipment including a grab bucket or a grab bucket suspended from a crane boom head. Also known as grab dredger.

Gradeability: The performance of earthmovers on

various inclines, measured in percent grade.

Grader: High-body wheeled vehicle with a leveling knife between the front and rear wheels; Used for fine grading on relatively loose and level ground.

Graetz number: A dimensionless number used in the study of streamline flow, equal to the mass flow rate of a fluid times its specific heat at constant pressure divided by the product of its thermal conductivity and a characteristic length.

Grain: A grain is a small, hard, dry seed with or without an attached hull or fruit layer harvested for human or animal consumption.

Graining: Simulating a grain such as wood or marble on a painted surface by applying a translucent stain, then working it into suitable patterns with tools such as special combs, brushes, and rags.

Gram: The gram is a metric system unit of mass.

Gram centimeter: A unit of energy in the centimetre gram second gravitational system

Gram force: A unit of force in the centimeter-gram-second gravitational system, equal to the gravitational

force on a 1 gram mass at a specified location.

Graphical statics: A method for determining the forces acting on a solid in a state of equilibrium, in which the forces are depicted on the diagram by straight lines, the length of which is proportional to the magnitude of the forces.

Grasshopper linkage: A straight line method used in some early steam engines.

Gravel pump: A centrifugal pump with renewable impellers and lining, used to pump a mixture of gravel and water.

Gravimetry: Measurement of gravitational force.

Gravitational constant: The constant of proportionality in Newton's law of gravitation, equal to the gravitational force between any two particles times the square of the distance between them, divided by the product of their masses.

Gravitational displacement: The gravitational field strength times the gravitational constant.

Gravitational field: A field in a region of space in which a test particle will experience a gravitational force, quantifying the gravitational force per unit mass acting on the particle at a specific point.

Gravitational force: The force on a particle due to its gravitational attraction to other particles.

Gravitational instability: Instability of a dynamic system in which gravity is the restoring force.

Gravitational potential: The amount of work that must be done against gravitational forces to move a particle of unit mass to a specified position from its original position, usually a point at infinity.

Gravitational potential energy: The energy that a system of particles has by virtue of their positions

Gravitational systems of units: Systems in which length, force and time are considered fundamental, and the unit of force is the force of gravity acting on a standard body at a specific location on the earth's surface.

Gravity: The gravitational attraction at the surface of a planet or other celestial body.

Gravity conveyor: Any non-driven conveyor, such as a gravity chute or roller con-

veyor, that uses gravity to move materials in a downward path.

Gravity vector: The force of gravity per unit mass at a given point.

Gravity wheel conveyor: Downward sloping conveyor chute with closely spaced wheel blocks on the axles that move containers or flat-bottomed objects from point to point by gravity.

Gray body: An energy radiator which has a blackbody energy distribution, reduced by a constant factor, throughout the radiation spectrum or within a certain wavelength interval.

Griffith's criterion: A criterion for the fracture of a brittle material under biaxial stress, based on the theory that the strength of such a material is limited by small cracks.

Griffiths' method: A method of measuring mechanical heat equivalent in which the rise in temperature of a known mass of water is compared with the electrical energy required to produce that rise.

Grinder: Any device or machine that grinds, such as a pulverizer or a grinding wheel.

Grinding burn: Overheating of the localized work area during grinding.

Grinding medium: Any material including balls and rods, used in a grinding mill.

Grinding mill: A machine consisting of a rotating cylindrical drum, that reduces the size of particles of ore or other materials fed into it, three main types are ball, rod, and tube mills are there.

Grinding ratio: Ratio of the volume of ground material removed from the work piece to the volume removed from the grinding wheel.

Grinding stress: Residual tensile or compressive stress, or a combination of both, on the surface of a material due to grinding

Grindstone: A stone disk on a revolving axle, used for grinding, smoothing, and shaping.

Grizzly crusher: A machine with a series of parallel rods or bars for crushing rock and sorting particles by size.

Groover: A tool for forming grooves in a slab of concrete not yet hardened.

Grooving saw: A circular saw for cutting grooves.

Grouting: The act or process of applying grout or of injecting grout into grout holes or crevices of a rock.

Guide bearing: A sleeve bearing used to guide a machine element in its longitudinal motion, usually without rotating the element.

Guide idler: An intermediate roller with a support structure mounted on the conveyor frame to guide the belt along a predetermined horizontal path, usually by contacting the edge of the belt.

Guide pin: A pin used to line up a tool or die with the work.

Guides: Pulleys to guide the drive belt or rope in a new direction or to keep it from the desired direction.

Guidewire: A wire embedded in the surface of the path travelled by an electromagnetically guided automated guided vehicle.

Gukhman number: Dimensionless number used in the study of convective heat transfer during evaporation,

Gun reaction: The force applied to the gun mount due to the movement of the gun backward as a result of the forward movement of the projectile and hot gases.

Gust load: The wind load on an antenna due to gusts.

Guy: A rope or wire securing a pole, derrick, or similar temporary structure in a vertical position.

Guy derrick: A derrick with a vertical pole supported by guy ropes to which a boom is attached by rope or cable suspension at the top and by a pivot at the foot.

Gyratory crusher: Machine for primary destruction in the form of two cones, an external fixed cone and a solid internal vertical cone, mounted on an eccentric support. Also known as a rotation breaker.

Gyratory screen: Box machine with a row of horizontal meshes nested in a vertical stack with decreasing mesh openings; an almost circular motion leads to the fact that the substandard material is sieved through each sieve sequentially

Gyro dynamics: The study of rotating bodies, especially those subject to precession.

Gyro pendulum: A gravity pendulum attached to a rapidly spinning gyro wheel.

Gyroscopic couple: The turning moment which resist any change of the inclination of the axis of rotation of a gyroscope.

Gyroscopic precession: Rotation of the gyroscope's axis of rotation under the action of an external torque acting on the gyroscope; the axis always turns in the direction of the torque.

Gyroscopics: The branch of mechanics anxious with gyroscopes and their use in stabilization and control of ships, aircraft, projectiles, and other objects.

Gyro wheel: The speedily spinning wheel in a gyroscope, which resists being disturbed.

Hamilton-Jacobi theory

Hamilton-Jacobi theory: A theory that provides a means for discussing the motion of a dynamical system in terms of a single first-order partial differential equation.

Hamilton's principle: The variational principle, which states that the path of a conservative system in the configuration space between two configurations is such that the integral of the Lagrange function in time is a minimum or maximum relative to the nearest paths between the same endpoints and takes the same time.

Hammer drill: Any of the three types of compressed air rock drills (downhole, lead, and stop), in which the hammer strikes rapidly against a freely held piston while the bit remains at the bottom of the hole, bouncing off the bit. slightly on every hit, but doesn't reciprocate.

Hammer mill: A type of impact mill or crusher in which materials are reduced in size by hammers rotating rapidly in a vertical plane inside a steel housing. Also known as bit mill.

Hammer milling: Crushing or fracturing materials in a hammer mill.

Hand brake: A manually operated brake.

Handcar: A small four-wheeled carriage with a hand pump used on railway tracks to transport workers and equipment for construction or repair work; other cars for the same purposes are motor vehicles.

Handle: The arm connecting the bucket with the boom in a dipper shovel or hoe.

Hanging drop atomizer: A spray device used in gravity atomization; functions due to the quasi-static ejection of a drop from a wetted surface.

Hanging load: A load that can be suspended from a hoisting line or hook device on a drill tripod or rig without causing bending of the rig or tripod members.

Hardinge feeder-weigher: A pivoted, short belt conveyor which controls the rate of material flow from a hopper by weight per cubic foot.

Hardinge mill: A tricone type of ball mill; the cones become steeper from the feed end toward the discharge end.

Hardwood bearing: Liquid film bearing made of lignum vitae with natural resin or hard maple impregnated with oil, grease or wax.

Harmonic drive: Drive system that uses internal and external gears to provide smooth motion.

Harmonic motion: simple harmonic motion is a special type of periodic motion where the restoring force on the moving object is directly proportional to the magnitude of the object's displacement and acts towards the object's equilibrium position

Harmonic speed changer: A mechanical drive system used to transmit rotary, linear or angular motion with high gear ratios and positive motion.

Harmonic synthesizer: A machine which combines elementary harmonic constituents into a single periodic function; a tide-predicting machine is an example.

Hartford loop: Condensate return device for low pressure steam heating systems with a stable water supply to the boiler.

Hayward grab bucket: Clamshell bucket used for handling coal, sand, gravel and other bulk materials.

Hayward orange peel: Clamshell bucket that works like a grab bucket, but with four blades that can be rotated to close.

Headache post: A stand mounted on a wireline drilling rig to support the end of the rocker arm when the rig is not running.

Head meter: A flow meter that is dependent upon change of pressure head to operate.

Head motion: A vibrator on a table reciprocating hub that transmits motion to the deck.

Head pulley: The pulley at the discharge end of a conveyor belt; may be either an idler or a drive pulley.

Head-pulley-drive conveyor: A conveyor having the belt driven by the head pulley without a snub pulley.

Head section: That element of belt conveyor which consists of a drive pulley, a head pulley which may or may not be a drive pulley, belt idlers if included, and the necessary framing.

Head shaft: The shaft is driven by a chain and is mounted on the discharge end of the chain conveyor; it supports the sprocket that drives the brake chain.

Headstock: The device on a lathe for carrying the revolving spindle.

Heat: It is the form of Energy. Which transfer due to a temperature difference between the source from which the energy is coming and a sink toward which the energy is going.

Heat balance: The equilibrium which is known to exist when all sources of heat gain and loss for a given region or body are accounted for.

Heat budget: A statement of the total heat inflow and outflow for a planet, spacecraft, biological organism, or other object.

Heat capacity: The amount of heat required to raise the temperature of the system one degree in a specific manner, usually at constant pressure or constant volume. Also known as heat output.

Heat conduction: The flow of thermal energy by microscopic collisions of particles and movement of electrons within a body. The colliding particles, which include molecules, atoms and electrons,.

Heat convection: Convection is the transfer of heat from one place to another due to the movement of fluid.

Heat death: The state of any isolated system, when its entropy reaches its maximum, in which matter is completely disordered and has a uniform temperature, and energy is not available to perform work.

Heat engine: A machine that converts heat into work (mechanical energy).

Heater: An electric heating element for supplying heat to the cathode with indirect heating in a vacuum tube. Also known as vacuum tube heater.

Heat flow (Transfer): Heat transfer is a discipline of thermal engineering that concerns the generation, use, conversion, and exchange of thermal energy between physical systems.

Heat flux: The quantity of heat transferred across a surface of unit area in a unit time. Also known as thermal flux.

Heat of ablation: A measure of the effective heat capacity of an ablating material, numerically the heating rate input divided by the mass loss rate which results from ablation.

Heat of adsorption: Increase in enthalpy when 1 mole of a substance is adsorbed on another at constant pressure.

Heat of aggregation: The

augment in enthalpy when an aggregate of matter, such as a crystal, is formed at constant pressure.

Heat of compression: Heat generated when air is compressed.

Heat of condensation: The increase in enthalpy accompanying the conversion of 1 mole of vapor into liquid at constant pressure and temperature.

Heat of crystallization: The increase in enthalpy when 1 mole of a matter is transformed into its crystalline state at constant pressure.

Heat of fusion: An increase in enthalpy accompanying the transformation of 1 mole or unit mass of a solid into a liquid at its melting point at constant pressure and temperature. Also known as latent heat of fusion.

Heat of mixing: The difference between the enthalpy of a mixture and the sum of the enthalpies of its components at the same pressure and temperature.

Heat rate: Heat rate is **one measure of the efficiency of electrical generators/power plants** that convert a fuel into heat and into electricity. The

heat rate is the amount of energy used by an electrical generator/power plant to generate one kilo watt hour (kWh) of electricity.

Heat release: The quantity of heat released by a furnace or other heating mechanism per second, divided by its volume.

Heat transfer coefficient: The amount of heat that passes through a unit of area of a medium or system per unit of time when the temperature difference between the boundaries of the system is 1 degree.

Heat transport: Process by which heat is carried past a fixed point or across a fixed plane, as in a warm current.

Heat wheel: In a ventilation system, a device for conditioning the incoming air, forcing it to approach thermal equilibrium with the outgoing air. hot incoming air is cooled and cold incoming air is heated.

Heavy-duty car: A railway motorcar weighing more than 1400 pounds (635 kilo-grams),

Heavy section car: A railway motorcar weighing 1200-1400 pounds (544-635 kilograms) and propelled by an

8-12 horsepower (6000-8900 watts) engine.

Hectare: The hectare is a non-SI metric unit of area equal to a square with 100-metre sides, or 10,000 m², and is primarily used in the measurement of land.

Hectogram: A unit of mass equal to 100 grams.

Hectolitre: A metric unit of volume equal to 100 liters or to 0.1 cubic meter. Abbreviated hl.

Hectometre: A unit of length equal to 100 meters.

Heel block: A block or plate that is usually fixed on the die shoe to minimize deflection of a punch or cam.

Helical angle: In the study of torsion, the angular displacement of a longitudinal element, initially straight on the surface of the untwisted bar, which becomes spiral after twisting.

Helical conveyor: Conveyor for conveying bulk materials, consisting of a horizontal shaft with spiral blades or belts rotating inside a stationary tube.

Helical-flow turbine: A steam turbine in which the steam is directed tangentially and radially inward by noz-

zles against buckets milled in the wheel rim; the steam flows in a helical path, re-entering the buckets one or more times.

Helical gear: Gear wheels running on parallel axes, with teeth twisted oblique to the gear axis.

Helical milling: Milling in which the work is simultaneously rotated and translated.

Helical spline broach: A broach used to produce internal helical splines having a straight sided or involute form.

Helium refrigerator: A refrigerator which uses liquid helium to cool substances to temperatures of 4 K or less.

Hemming: Forming of an edge by bending the metal back on itself.

Hereditary mechanics: A field of mechanics in which quantities such as stress depend not only on other quantities, such as deformation, at the same time, but also on integrals that include the values of such quantities at previous times.

Herpolhode: A curve drawn on a fixed plane by the point of contact of the plane with the ellipsoid of inertia of a rotating rigid body not subject to external torque.

Herringbone gear: The equivalent of two helical gears of opposite hand placed side by side.

Hertz's law: A law that determines the radius of contact between a sphere of elastic material and a surface through the radius of the sphere, the normal force acting on the sphere, and Young's modulus for the material of the sphere.

Heterogeneous strain: A deformation in which the offset components of a body point cannot be expressed as linear functions of the original coordinates.

Heteromorphic transformation: Changes in the values of thermodynamic variables of a system in which one or more components of a substance also undergo a change in state.

High-efficiency particulate air filter: An air filter capable of reducing the concentration of solid particles (0.3 millimeter in diameter or larger) in the airstream by 99.97%. Also known as HEPA filter.

Higher pair: A link in a mechanism in which the mating parts have surface (instead of line or point) contact.

High front shovel: A power shovel with a dipper stick mounted high on the boom for stripping and overburden removal.

High heat: Heat absorbed by the cooling medium in a calorimeter when products of combustion are cooled to the initial atmospheric (ambient) temperature.

High intensity atomizer: A type of atom used in electrostatic atomization, based on sufficient pressure to remove the tensile strength of the liquid.

High lift truck: A forklift truck with a fixed or telescoping mast to permit high elevation of a load.

High speed machine: A diamond drill capable of rotating a drill string at a minimum of 2500 revolutions per minute, as contrasted with the normal maximum speed of 1600-1800 revolutions per minute attained by the average diamond drill.

High temperature water boiler: A boiler which provides hot water, under pressure, for space heating of large areas.

Hildebrand function: The heat of vaporization of a

compound as a function of the molal concentration of the vapor; it is nearly the same for many compounds.

Hill climbing: Adjustment, either continuous or periodic of a self-regulating system to achieve optimum performance.

Hobbing machine: A machine for cutting gear teeth in gear blanks or for cutting worm, spur, or helical gears. Also known as hobber.

Hoe shovel: A revolving shovel with a pull type bucket rigidly attached to a stick hinged on the end of a live boom.

Hoist: To move or lift something by a rope and pulley device.

Hoist back-out switch: A protective switch that permits hoist operation only in the reverse direction in case of over wind.

Hoist cable: A fibrous rope, cable or chain that acts on the pulleys and pulleys of the hoisting machine.

Hoisting: Raising a load, especially by means of tackle.

Hoisting machine: Mechanism for lifting and lowering material with intermittent motion, keeping material freely suspended.

Hoisting power: The capacity of the hoisting mechanism on a hoisting machine.

Hoist over speed device: A device used to prevent the winch from operating at speeds in excess of the set speed by activating the emergency brake when the set speed is exceeded.

Hoist overwind device: A device that can activate an emergency brake when a lifted load moves outside a predetermined point into a hazardous area.

Hoist slack brake switch: A device that automatically cuts off power to the winch motor and applies the brake if the brake links need to be tightened or if the brakes need to be relined.

Hoistway: A shaft for one or more elevators, lifts, or dumbwaiters.

Hold back: A brake on an inclined belt conveyor system that is automatically activated in the event of a power failure, thus preventing the loaded belt from moving downwards.

Hollander: An elongate tube with a central mid-feather and a cylindrical beater roll; formerly used for stock preparation in paper manufacture.

hollow mill: A milling cutter with three or more cutting edges that revolve around the cylindrical work piece.

Hollow-rod churn drill: A churn drill with hollow rods instead of steel wire rope.

Hollow shafting: A shaft made from hollow rods or hollow tubes to minimize weight, provide internal support, or allow another shaft to pass through the interior.

Holonomic constraints: An integrable system of differential equations describing the constraints on the motion of the system; a function that binds several variables,

Holonomic system: A system in which the constraints are such that the original co-ordinates can be expressed in terms of independent coordinates and possibly also the time.

Holzer's method: A method for determining the forms and frequencies of torsional vibrations of a system, in which the system consists of several flywheels on a massless flexible shaft and, starting with a test frequency and move-

ment for one flywheel, determines the torques and movements of successive flywheels.

Homenergic flow: Fluid flow in which the sum of kinetic energy, potential energy, and enthalpy per unit mass is the same at all locations in the fluid and at all times.

Homogeneous strain: A deformation in which the displacement components of any point on the body are linear functions of the original coordinates.

Homogenizer: A machine that mixes or emulsifies a substance by pushing it through fine holes onto a hard surface.

Homomorphous transformation: Change in the values of thermodynamic variables of a system in which none of the constituent substances undergoes a change in state.

Hone: A machine for honing that consists of a holding device containing several oblong stones arranged in a circular pattern.

Honeycomb radiator: A heat exchange appliance utilizing many small cells, shaped like a bees' comb, for cooling circulating water in an automobile.

Honing: The process of removing a relatively small amount of material from a cylindrical surface by means of abrasive stones to obtain a desired finish or extremely close dimensional tolerance.

Hookean deformation: Deformation of a body which is proportional to the force applied to it.

Hookean solid : An ideal solid which obeys Hooke's law exactly for all values of stress, however large.

Hooke's joint: A simple universal joint; consists of two yokes attached to their respective shafts and connected by means of a spider. Also known as Cardan joint.

Hooke's law: The law that the stress of a solid is directly proportional to the strain applied to it.

Horizontal auger: A rotary drill, usually powered by a gasoline engine, for making horizontal blasting holes in quarries and open-cast pits.

Horizontal boiler: A water tube boiler having a main bank of straight tubes inclined toward the rear at an angle of 5 to 15 from the horizontal.

Horizontal boring machine: A boring machine adapted for work not conveniently revolved, for milling, slotting, drilling, tapping, boring, and reaming long holes and for making interchangeable parts that must be produced without jigs and fixtures.

Horizontal broaching machine: A pull type broaching machine having the broach mounted on the horizontal plane.

Horizontal crusher: Rotary size reducer in which the crushing cone is supported on a horizontal shaft, needs less headroom than vertical models.

Horizontal drilling machine: A drilling machine in which the drills run horizontally.

Horizontal engine: An engine with horizontal stroke

Horizontal firing: Combustion of fuel in the boiler furnace, in which the burners discharge fuel and air into the furnace horizontally.

Horizontal lathe: A horizontally mounted lathe, with which longitudinal and radial movements are processed by a rotating part.

Horizontal milling ma-

chine: Knee milling machine with horizontal spindle and rotary table for cutting spirals.

Horizontal pendulum: A pendulum that move in a horizontal plane, such as a compass needle turning on its pivot.

Horizontal return tubular boiler: Fire-tube boiler having cylindrical shell pipes attached to end gates; combustion products are transported under the lower half of the shell and back through pipes.

Horizontal-tube evaporator: A horizontally mounted shell and tube liquid evaporator most commonly used for preparing boiler feed water.

Horsepower: The unit of power in the British engineering system, equal to 550 foot-pounds per second, approximately 745.7 watts. Abbreviated hp.

Hostile-environment machine: A robot capable of operating in extreme conditions of temperature, vibration, humidity, pollution, electromagnetic or nuclear radiation.

Hot-air engine: A heat engine that uses air or other gases such as hydrogen, helium or nitrogen as a propellant and operates on cycles such as Stirling or Ericsson.

Hot air furnace: An encased heating element providing warm air to ducts for circulation by gravity convection or by fans.

Hot bulb: Pertaining to an ignition method used in semi diesel engines in which the fuel mixture is ignited in a separate chamber kept above the ignition temperature by the heat of compression.

Hotchkiss drive: An automobile rear suspension designed to take torque reactions through longitudinal leaf springs.

Hot saw: A power saw used to cut hot metal.

Hot-water heating: Heating system for a building in which the heating medium is hot water and the heat transfer media are radiators, convectors or panel coils. Also known as hot water heating.

Hot well: A chamber for collecting condensate, as in a steam condenser serving an engine or turbine.

Hour: A unit of time equal to 3600 seconds.

Humidifier: A device for

humidifying the air and maintaining the desired humidity level.

Humphrey gas pump: Combined internal combustion engine and pump, in which the metal piston is replaced by a column of water.

Humphries equation: An equation which gives the ratio of specific heats at constant pressure and constant volume in moist air as a function of water vapor pressure.

Huttig equation: An equation which states that the ratio of the volume of gas adsorbed on the surface of a nonporous solid at a given pressure and temperature to the volume of gas required to cover the surface completely with a unimolecular layer

Hydraulic accumulator: A hydraulic flywheel that stores potential energy by storing a certain amount of hydraulic fluid under pressure in a suitable closed vessel.

Hydraulic actuator: A cylindrical or hydraulic motor that converts hydraulic energy into useful mechanical work; the mechanical movement produced can be linear, rotary, or oscillatory.

Hydraulic air compressor: A device in which water falling

through a pipe entrains air, which is released under compression to perform useful work.

Hydraulic backhoe: A backhoe operated by a hydraulic mechanism.

Hydraulic brake: A brake in which the retarding force is applied through the action of a hydraulic press.

Hydraulic circuit: A circuit whose operation is similar to that of an electrical circuit, except that electrical currents are replaced by currents of water or other fluids, as in hydraulic control.

Hydraulic classifier: A classifier in which particles are sorted by their specific gravity in a hydraulic water stream that rises at a controlled speed; heavier particles are pulled down and thrown to the bottom

Hydraulic conveyor: A system for handling material such as ash from a coal stove; Debris is flushed from a hopper or slag tank into a grinder, which is discharged into a pump for transport to a landfill site or to a dewatering hopper.

Hydraulic cylinder: The cylindrical chamber of a positive displacement pump.

Hydraulic dredge: Dredger consisting of a large suction pipe that is mounted on the body and supported and moved by an arrow, a mechanical stirrer or cutting head that churns the earth in front of the pipe, and centrifugal pumps mounted on the dredger that suck in water. and bulk materials.

Hydraulic drill: A hydrodynamically driven rotary drill is used to drill holes in coal or rocks, or to drill holes.

Hydraulic drive: A mechanism that transfers motion from one shaft to another, and the ratio of the speeds of the shafts is regulated by hydrostatic or hydrodynamic means.

Hydraulic elevator: An elevator operated by water pressure. Also known as hydraulic lift.

Hydraulic excavator digger: An earthmoving machine that uses hydraulic pistons to drive mechanical digging elements.

Hydraulic jack: A jack in which force is applied through the mechanism of a hydraulic press.

Hydraulic machine: A machine driven by a motor driven by a restricted flow of fluid such as oil or water under pressure.

Hydraulic motor: A motor activated by water or other liquid under pressure.

Hydraulic nozzle: An atomizing device in which fluid pressure is converted into fluid velocity.

Hydraulic power system: Power transmission system, consisting of mechanisms and auxiliary components that generate, transmit, control and use hydraulic energy.

Hydraulic press: A combination of large and small cylinders connected by a pipe and filled with liquid, so that the pressure of the liquid created by a small force on the piston of the small cylinder results in a large force on the large piston.

Hydraulic ram: A device for pumping running water to a higher level by using the kinetic energy of the flow; the flow of water in the supply line periodically stops, so that a small part of the water rises by the high-speed pressure of the greater part. Also known as hydraulic pump.

Hydraulic rope geared elevator: The lift is lifted using a system of ropes and pulleys attached to a piston in the hydraulic cylinder.

Hydraulic scale: An industrial scale in which the load applied to the load-cell piston is converted to hydraulic pressure.

Hydraulic separation: Mechanical classification using a hydraulic classifier

Hydraulic shovel: Swing shovel in which cylinders or motors replace drums and cables.

Hydraulic sprayer: A machine that sprays large quantities of insecticide or fungicide on crops.

Hydraulic stacker: A tiering machine whose carriage is raised or lowered by a hydraulic cylinder.

Hydraulic swivel head: In a drill machine, a swivel head equipped with hydraulically actuated cylinders and pistons to exert pressure on and move the drill rod string longitudinally.

Hydraulic turbine: A water turbine is a rotary machine that converts kinetic energy and potential energy of water into mechanical work. .

Hydro cyclone: A cyclone separator in which solids are removed from the water stream and classified by centrifugal force.

Hydroelectric generator: An electrical rotating machine that converts the mechanical energy of a hydraulic turbine or water wheel into electrical energy.

Hydroelectric plant: A facility where electricity is generated by hydroelectric generators. Also known as a hydroelectric power plant.

Hydrometer: A direct-reading device for indicating the density, specific gravity, or some similar characteristic of liquids.

Hydroseparator: A separator in which solids in suspension are agitated by hydraulic pressure or stirring devices.

Hydrostatic balance: An equal arm balance in which an object is weighed first in air and then in a beaker of water to determine its specific gravity.

Hydrostatic bearing: A sleeve bearing in which high pressure oil is pumped into the area between the shaft and the bearing so that the shaft is raised and supported by an oil film.

Hydrostatic roller conveyor: Part of a roller conveyor with rollers weighted with fluid to control the speed of moving objects.

Hydrostatic strength: The ability of a body to withstand hydrostatic stress.

Hydrostatic stress: A state in which there are equal compressive stresses or equal tensile stresses in all directions and the absence of shear stresses in any plane.

Hyperoid axle: A type of rear axle drive gear set which generally carries the pinion 1.5-2 inches (38-51 millimeters) or more below the centerline of the gear.

Hypoid gear: Gear wheels connect hysteresis in nonparallel, nonintersecting shafts, usually at right angles.

Hypoid generator: A gear cutting machine for making hypoid gears.

Hysteresis clutch: A clutch in which torque is generated by attraction between the induced poles in the magnetized iron ring and the driving field.

Hysteresis damping: Vibration damping due to energy loss due to mechanical hysteresis.

Hysteretic damping: Damping of an oscillatory system in which the braking force is proportional to the speed and inversely proportional to the frequency of vibration.

I beam: Rolled iron or steel joist with I section and short flange.

Ice line: Graph of freezing point of water as a function of pressure.

Ideal gas: Also identified as perfect gas. A gas whose molecules are infinitely small and exert no force on each other.

Ideal gas law: The equation of state of an ideal gas which is a good approximation to real gases at sufficiently high temperatures and low pressures; that is, $PV = mRT$, where P is the pressure, V is the volume, T is the temperature, and R is the gas constant. m is mass.

Idle: To run without a load.

Idler arm: In an automotive steering system, a link that supports the tie rod and transmits steering motion to both wheels through the ends of the tie rod.

idler gear : A gear placed between a driving gear and a driven gear to transfer motion, without any change of direction or of gear ratio.

Idler pulley: A pulley used to direct and tighten the belt or chain of a conveyor system.

Idler wheel: A wheel used to convey motion or to guide and support something.

Idle-stop solenoid: An electrically operated plunger in a carburetor that supply a predetermined throttle setting at idle and closes the throttle completely when the ignition switch is turned off.

Idling jet: A carburetor part that initiate gasoline during minimum load or speed of the engine.

Idling system: A system to attain adequate metering forces at low air speed and small throttle openings in an automobile carburetor in the idling position.

Ignition lag: In the internal combustion engine, the time gap between the passage of the spark and the inflammation of the air-fuel mixture.

Ignition system: The arrangement in an internal combustion engine that initiates the chemical reaction between fuel and air in the cylinder charge by producing a spark.

I-head cylinder: The internal combustion engine construction having both inlet and exhaust valves located in the cylinder head.

Immersion scanning: An

ultrasonic scan that immerses both the transducer and the object to be scanned in water or other liquid to provide good coupling while the transducer moves around the object.

Impact: A strong collision between two objects sufficient to cause a significant change in the momentum of the system in which it acts.

Impact area: An area with designated boundaries in which all objects that travel over a range are to make contact with the ground.

Impact breaker: A device that utilizes the energy from falling stones in addition to power from massive impellers for complete breaking up of stone.

impact crusher: A machine for crushing large chunks of solid materials by sharp blows imposed by rotating hammers, or steel plates or bars; some crushers accept lumps as large as 28 inches (about 70 centimeters) in diameter, reducing them to 1/4 inch (6 millimeters) and smaller.

Impact energy: The energy required to fracture a material. Also known as impact strength.

Impact grinding: A method used to break up particles by direct fall of crushing bodies on them.

Impact mill: A unit that decrease the size of rocks and minerals by the action of rotating blades projecting the material against steel plates.

Impact roll: An idler roll protected by a covering of a resilient material from the shock of the loading of material onto a conveyor belt, so as to reduce the damage to the belt.

Impact screen: A screen designed to swing or lock forward on load and stop suddenly when it touches a stop.

Impact strength: Ability of a material to resist shock loading.

Impact stress: Force per unit area imposed on a material by a suddenly applied force.

Impact velocity: The velocity of a projectile or missile at the instant of impact. Also known as striking velocity.

Impact wrench: A compressed-air or electrically operated wrench that gives a rapid succession of sudden torques.

Impeller: The rotating part of a turbine, blower, fan, axial or centrifugal pump, or mixing apparatus.

Impeller pump: Any pump with a mechanical agency to provide continuous power to move liquids.

Impulse: The integral of a force over an interval of time.

Impulse turbine: A prime mover where fluid under pressure enters a stationary nozzle where its pressure (potential) energy is converted to velocity (kinetic) energy and absorbed by the rotor.

Inch: A unit of length in common use in the United States and the United Kingdom, equal to 1/12 foot or 2.54 centimeters.

Inclined cableway: A monocable arrangement in which the track cable has a slope sufficiently steep to allow the carrier to run down under its own weight.

Inclined plane: A plane surface at an angle to some force or reference line.

Incomplete lubrication: Lubrication that takes place when the load on the rubbing surfaces is carried partly by a fluid viscous film and partly by areas of boundary lubrication; friction is intermediate among

that of fluid and boundary lubrication.

Incompressibility: Quality of a matter which maintains its original volume under increased pressure.

Independent suspension: In automobiles, a system of springs and guide links where the wheels are mounted independently on the chassis.

Index center: One of two machine tool centers used to hold work and to rotate it by a fixed amount.

Index chart: A chart used in combination with an indexing or dividing head, which correlates the index plate, hole circle, and index crank motion with the required angular subdivisions.

Index crank: The crank handle of an index head used to rotate the spindle.

Index head: A headstock that can be affixed to the table of a milling machine, planer, or shaper; work may be mounted on it by a chuck or centers, for indexing.

Indexing: The process of providing discrete spaces, parts, or angles in a workpiece by using an index head.

Indexing fixture: A fixture that changes position with regular steplike movements.

Indicated horsepower: The horsepower delivered by an engine as calculated from the average pressure of the working fluid in the cylinders and the displacement.

Induced draft: A mechanical draft formed by suction stream jets or fans at the point where air or gases leave a unit.

Induced-draft cooling tower: An arrangement for cooling water by circulating air where the load is on the suction side of the fan.

Induction pump: Any pump worked by electromagnetic induction.

Industrial engineering: A branch of engineering anxious with the design, improvement, and installation of integrated systems of people, materials, and equipment.

Inelastic: Not competent of sustaining a deformation without permanent change in size or shape.

Inelastic buckling: Unexpected increase of deflection or twist in a column when compressive stress reaches the elastic limit but before elastic buckling develops.

Inelastic collision: A collision in which the total kinetic energy of the colliding particles is not the same after the collision as before it.

Inelastic stress: A force acting on a solid which create a deformation such that the original shape and size of the solid are not restored after removal of the force.

Inertia ellipsoid: An ellipsoid used in describing the motion of a rigid body; it is fixed in the body, and the distance from its center to its surface in any direction is inversely proportional to the square root of the moment of inertia about the corresponding axis.

Inertia governor: A speed-control tool utilizing suspended masses that respond to speed changes by reason of their inertia.

Inertial force: The fictitious force acting on a body as a result of using a non-inertial frame of reference; examples are the centrifugal and Coriolis forces that appear in rotating coordinate systems.

Inertial mass: The mass of a substance as determined by Newton's second law, in contrast to the mass as determined by the proportionality to the gravitational force.

Inertial reference frame: A coordinate arrangement in which a body moves with constant velocity as long as no force is acting on it. Also known as inertial coordinate system.

Inertia matrix: A matrix M used to express the kinetic energy T of a mechanical system during small displacements from an equilibrium position

Inertia starter: An apparatus utilizing inertial principles to start the rotator of an internal combustion engine.

Inertia tensor: A tensor connected with a rigid body whose product with the body's rotation vector yields the body's angular momentum

Inextensional deformation: A bending of a surface that leaves unchanged the length of any line drawn on the surface and the curvature of the surface at each point.

In-feed centerless grinding: A metal-cutting process by which a cylindrical workpiece is ground to a prescribed surface smoothness and diameter by the insertion of the workpiece between a grinding wheel and a canted regulating wheel; the rotation of the regulating wheel controls the rotation

and feed rate of the workpiece.

Influence line: A graph of the shear, stress, bending moment, or other outcome of a movable load on a structural member versus the position of the load.

Inhaul cable: In a cable excavator, the line that drag the bucket to dig and bring in soil.

Inherent damping: A process of vibration damping which makes use of the mechanical hysteresis of such materials as rubber, felt, and cork.

Injection carburetor: A carburettor in which fuel is transported under pressure into a heated part of the engine intake system. Also known as pressure carburetor.

Injection pump: A pump that forces a measured amount of fuel through a fuel line and atomizing nozzle in the combustion chamber of an internal combustion engine.

Inlet box: A closure at the fan inlet or inlets in a boiler for attachment of the fan to the duct system.

Inlet valve: The valve through which a fluid is haggard into the cylinder of a positive-displacement engine, pump, or compressor.

In-line engine: A multiple-cylinder engine with cylinders aligned in a row.

In-line linkage: A power steering connection which has the control valve and actuator combined in a single assembly.

Instantaneous axis: The axis around which a rigid body is carrying out a pure rotation at a given instant in time.

Instantaneous center: A point about which a rigid body is rotating at a given instant in time.

Instantaneous recovery: The instant reduction in the strain of a solid when a stress is removed or reduced, in contrast to creep recovery.

Instantaneous strain: The instant deformation of a solid upon initial application of a stress, in contrast to creep strain.

Instrument system: An arrangement which integrates one or more instruments with auxiliary or associated devices for detection, observation, measurement, automatic control, automatic computation, communication, or data processing.

Intake manifold: An arrangement of pipes which feeds fuel to the various cylinders of a multi cylinder internal combustion engine.

Intake stroke: The fluid admission phase or travel of a reciprocating piston and cylinder mechanism as, for example, in an engine, pump, or compressor.

Intake valve: The valve which opens to permit air or an air-fuel mixture to enter an engine cylinder.

Integrable system: A dynamical arrangement whose motion is governed by an integrable differential equation.

Integral-furnace boiler: A kind of steam boiler which incorporates furnace water-cooling in the circulatory system.

Intercondenser: A condenser among stages of a multi-stage steam jet pump.

Intercooler: A heat exchanger for cooling fluid among stages of a multi-stage compressor with consequent saving in power.

Interface resistance: Impairment of heat stream caused by the imperfect contact between two materials at an interface.

Intermediate gear: An idler gear insert between a driver and driven gear.

Intermittent firing: Cyclic firing whereby fuel and air are burned in a furnace for frequent short time periods.

Internal brake: It is used in motor vehicles. A friction brake in which an internal shoe follows the inner surface of the rotating brake drum, wedging itself between the drum and the point at which it is anchored.

Internal broaching: Taking away of material on internal surfaces, by means of a tool with teeth of progressively increasing size moving in a straight line or other prescribed path over the surface, other than for the origination of a hole.

Internal combustion engine: It is the most common form of heat engines, which are used in vehicles, boats, ships, airplanes, and trains. It is named as the fuel is ignited in order to do work inside the engine.

Internal energy: It is denoted by (U). A characteristic property of the state of a thermodynamic system, introduced in the first law of thermodynamics.

Internal floating-head exchanger: Tube-and-shell heat exchanger in which the tube sheet (support for tubes) at one end of the tube bundle is free to move.

Internal force: Force exert by one part of a system on another.

Internal friction: Change of mechanical strain energy to heat.

Internal vibrator: A vibrating tool which is drawn vertically through placed concrete to achieve proper consolidation

Intertube burner: A burner which employ a nozzle that discharges between adjacent tubes.

Invariable line: A line which is parallel to the angular momentum vector of a body executing Poinsot motion, and which passes through the fixed point in the body about which there is no torque.

Invariable plane: A plane which is perpendicular to the angular momentum vector of a rotating rigid body not subject to external torque, and which is always tangent to its inertia ellipsoid.

Inverse cam: A cam that acts as a follower as a substitute of a driver.

Inverted engine: An engine where the cylinders are below the crankshaft

Involute spline broach: A broach that cuts several keys in the form of internal or external involute gear teeth.

Irreversible energy loss: Energy transformation method in which the resultant condition lacks the driving potential needed to reverse the process

Irreversible process: Irreversible processes are actual processes carried out in finite time with real substances. An example of an irreversible process is a spontaneous chemical reaction or electrochemical reaction.

Isenergic flow: Fluid flow in which the sum of the kinetic energy, potential energy, and enthalpy of any part of the fluid does not change as that part is carried along with the fluid.

Isenthalpic expansion: Expansion which takes place without any differ of enthalpy.

Isenthalpic process: A process that is carried out at constant enthalpy.

Isentrope: A line of equal or constant entropy.

Isentropic: Having constant entropy, at constant entropy. (S=Constant)

Isentropic compression: Compression which occurs without any change in entropy.

Isentropic expansion: Expansion which occurs without any change in entropy.

Isentropic flow: Flow of Fluid in which the entropy of any part of the fluid does not change as that part is carried along with the fluid.

Isentropic process: A process with constant entropy, a process which is both reversible and adiabatic.

Isobaric: constant pressure, with respect to either space or time.

Isobaric process: A thermodynamic process of a gas in which the heat transfer to or from the gaseous system causes a volume change at constant pressure. (P=Constant)

Isochronism: The property of having a uniform rate of operation or periodicity.

Isochronous governor: A governor that keeps the speed of a prime mover constant at all loads.

Isodynamic: Pertaining to equality of two or more forces or to constancy of a force.

Isometric process: A constant volume frictionless thermodynamic process in which the system is confined by mechanically rigid boundaries.

Isostatics: In photo elasticity studies of stress analyses, those curves, the tangents to which represent the progressive change in principal plane directions.

Isostatic surface: A surface in a three dimensional elastic body such that at each point of the surface one of the principal planes of stress at that point is tangent to the surface.

Isotherm: A curve or formula showing the liaison between two variables, such as pressure and volume, when the temperature is held constant.

Isothermal calorimeter: A calorimeter in which the heat received by a reservoir, containing a liquid in equilibrium with its solid at the melting point or with its vapor at the boiling point, is determined by the change in volume of the liquid.

Isothermal compression: Compression at constant temperature.

Isothermal equilibrium: The condition in which two or more systems are at the same temperature, so that no heat flows between them.

Isothermal expansion: Expansion of a substance while its temperature is held constant.

Isothermal flow: Flow of a gas in which its temperature does not change.

Isothermal layer: A layer of fluid, all points of which have the same temperature.

Isothermal magnetization: Magnetization of a substance held at constant temperature; used in combination with adiabatic demagnetization to produce temperatures close to absolute zero.

Isothermal process: Any constant temperature process, such as expansion or compression of a gas, accompanied by heat addition or removal from the system at a rate just adequate to maintain the constant temperature

Isothermal transformation : Any transformation of a matter which takes place at a constant temperature.

J **Jacket**: The gap around an engine cylinder through which a cooling liquid circulates.

Jack screw: A jack, that work by a screw mechanism.

Jack shaft: A counter shaft, particularly when used as an auxiliary shaft between two other shafts.

Jaeger-Steinwehr method: A refinement of the Griffiths process for determining the mechanical equivalent of heat, wherein a large mass of water, efficiently stirred, is used, the temperature rise of the water is small, and the temperature of the surroundings is carefully con-trolled.

Jaw clutch: A clutch that offers positive connection of one shaft with another by means of interlocking faces; may be square or spiral; the most common type of positive clutch.

Jaw crusher: A device for breaking rock between two steel jaws, one fixed and the other swinging.

Jeans viscosity equation: An equation which states that the viscosity of a gas is proportional to the temperature raised to a constant power, which is different for different gases.

Jeep: Four wheel-drive utility vehicle in wide use in all United States military services.

Jerk: The rate of change of acceleration;

Jerk pump: A pump that provides a precise amount of fuel to the fuel injection valve of an internal combustion engine at the time the valve opens; used for fuel injection.

Jet compressor: An apparatus that utilizing an actuating nozzle and a combining tube, for the pumping of a compressible fluid.

Jet condenser: A direct contact steam condenser utilizing the aspirating effect of a jet for the removal of non-condensable.

Jet drilling: A drilling process that utilizes a chopping bit, with a water jet run on a string of hollow drill rods, to chop through soils and wash the cuttings to the surface. Also known as wash boring.

Jet engine: Any engine that ejects a jet or stream of gas or fluid, obtaining all or most of its thrust by reaction to the ejection.

Jet mixer: A type of flow mixer or line mixer, depending on impingement of one liquid on the other to produce mixing.

Jet pump: A pump in which an accelerating jet entrains a second fluid to deliver it at elevated pressure.

J factor: A dimensionless equation used for the estimation of free convection heat transmission through fluid films.

Jib boom: An extension that is hinged to the upper end of a crane boom

Jib crane: Any of various cranes having a projecting arm

Jig back: An aerial ropeway with a pair of containers that move in opposite directions and are loaded or stopped alternately at opposite stations but do not pass around the terminals. Also known as reversible tramway; to-and-fro ropeway.

Jig borer: A device resembling a vertical milling machine designed for locating and drilling holes in jigs.

Jig grinder: A precision grinding machine used to locate and grind holes to size, especially in hardened steels and carbides.

Jig saw: A tool with a narrow blade suitable for cutting intricate curves and lines.

Jordan: A machine or engine used to refine paper pulp, consisting of a rotating cone, with cutters, that fits inside another cone, also with cutters.

Joule: It is equal to the energy transferred to an object when a force of one newton acts on that object in the direction of the force's motion through a distance of one metre.

Joule and Playfairs' experiment: An experiment in which the temperature of the maximum density of water is measured by taking the mean of the temperatures of water in two columns whose densities are determined to be equal from the absence of correction currents in a connecting trough.

Joule equivalent: The numerical relation between quantities of mechanical energy and heat.

Joule experiment: An experiment to observe intermolecular forces in a gas, in which one measures the heat absorbed when gas in a small vessel is allowed to expand into a second vessel which has been evacuated.

Joule's law: The law that at constant temperature the internal energy of a gas tends to a finite limit, independent of volume, as the pressure tends to zero.

Joule-Thomson coefficient: The ratio of the temperature change to the pressure change of a gas undergoing isenthalpic expansion.

Joule-Thomson effect: A change of temperature in a gas undergoing Joule-Thomson expansion.

Joule-Thomson expansion: The adiabatic, irreversible expansion of a fluid flowing through a porous plug or partially opened valve. Also known as Joule-Thomson process.

Joule-Thomson inversion temperature: A temperature at which the Joule-Thomson coefficient of a given gas changes sign.

Journal: That part of a shaft or crank which is supported by and turns in a bearing.

Journal bearing: A cylindrical bearing which wires a rotating cylindrical shaft

Journal friction: Friction of the axle in a journal bearing arising mostly from viscous sliding friction between journal and lubricant

Jumper tube: A short tube used to sidestep the flow of fluid in a boiler or tubular heater.

Junkers engine: A double opposed piston, two cycle internal combustion engine with intake and exhaust ports at opposite ends of the cylinder.

Kaplan turbine: A propeller type hydraulic turbine. Kaplan turbine is a complete reaction turbine that works based on the lift force generated on the impeller blades due to its aerofoil shape.

Kater's reversible pendulum: A gravity pendulum planned to measure the acceleration of gravity and consisting of a body with two knife edge supports on opposite sides of the center of mass.

Kauertz engine: A type of cat-and mouse rotary engine in which the pistons are vanes which are sections of a right circular cylinder.

Kellering: Three dimensional machining of a contoured surface by tracer milling. Also known as 3D milling.

Kelvin absolute temperature scale: Kelvin temperature scale, a temperature scale having an absolute zero below which temperatures do not exist. Absolute zero, or $0°K$, is the temperature at which molecular energy is a minimum.

Kelvin body: An ideal body whose shearing (tangential) stress is the sum of a term proportional to its deformation and a term pro-portional to the rate of change of its deformation with time. Also known as Voigt body.

Kelvin equation: The Kelvin equation describes the change in vapour pressure due to a curved liquid–vapor interface, such as the surface of a droplet. The vapor pressure at a convex curved surface is higher than that at a flat surface.

Kelvin scale: The kelvin is the base unit of temperature in the International System of Units, having the unit symbol K.

Kelvin temperature scale: An International Temperature Scale which agrees with the Kelvin absolute temperature scale within the limits of experimental determination.

Kennedy and Pancu circle: For a harmonic oscillator subject to hysteretic damping and subjected to a sinusoidally varying force, a plot of the in-phase and quadrature components of the displacement of the oscillator as the frequency of the applied vibration is varied.

Keyes equation: An equation of state of a gas which is designed to correct the van der Waals equation for the effect of surrounding molecules on the term representing the volume of a molecule.

Keyseater: A machine for milling beds or grooves in mechanical parts which receive keys.

Kickback: A backward thrust, such as the backward starting of an internal combustion engine as it is cranked or the reverse push of a piece of work as it is fed to a rotary saw.

Kickdown: Shifting to lower gear in an automotive vehicle.

Kick over: To start firing; applied to internal combustion engines.

Kick starter: A method for starting the operation of a motor by thrusting with the foot.

Kilobar: A unit of pressure equal to 1000 bars

Kilocalorie: Unit of heat energy equal to 1000 calories. Abbreviated kcal.

Kilogram: The unit of mass in the meter-kilogram-second system

Kilogram force: A unit of force equal to the weight of a 1-kilogram mass at a point on the earth's surface where the acceleration of gravity is 9.80665 m/s^2.

Kiloliter: A unit of volume equal to 1000 liters or to 1 cubic meter. Abbreviated kl.

Kilometer: A unit of length equal to 1000 meters. Abbreviated km.

Kinematically admissible motion: Any motion of a mechanical arrangement which is geometrically compatible with the constraints.

Kinematics: The study of the motion of a system of material particles without reference to the forces which act on the scheme.

Kinetic energy: The energy which a body possesses because of its motion; in classical mechanics, equal to one-half of the body's mass times the square of its speed.

Kinetic friction: The friction between two surfaces which are sliding over each other.

Kinetic momentum: The momentum which an element possesses because of its motion; in classical mechanics, equal to the particle's mass times its velocity.

Kinetic reaction: The negative of the mass of a body multiplied by its acceleration.

Kinetics: The dynamics of material bodies.

Kingpin: It is the main pivot in the steering mechanism of a car or other vehicle. The pin used between an automobile stub axle and an axle-beam.

Kip: A 1000-pound load.

Kirchhoff formula: A formula for the dependence of vapor pressure p on temperature T, valid over limited temperature ranges.

Kirchhoff's equations: In fluid dynamics, the Kirchhoff equations, named after Gustav Kirchhoff, describe the motion of a rigid body in an ideal fluid. An Equations which state that the partial derivative of the change of enthalpy (or of internal energy) during a reaction, with respect to temperature, at con-stant pressure (or volume) equals the change in heat capacity at constant pressure (or volume).

Kirchhoff vapor pressure formula: An estimated formula for the variation of vapor pressure p with temperature T, valid over a limited temperature range.

Kirkwood-Brinkely's theory: In terminal ballistics, a theory formulating the scaling laws from which the effect of blast at high altitudes may be inferred, based upon observed results at ground level.

Knee: In a knee-and-column type of milling machine, the part which supports the saddle and table and which can move vertically on the column.

Knee tool: A tool holder with a shape resembling a knee, such as the holder for simultaneous cutting and interval operations on a screw machine or turret lathe.

Knife-edge bearing: A balance beam or lever arm fulcrum in the form of a hardened steel wedge; used to reduce friction.

Knock-off: The usual stopping of a machine when it is operating improperly.

Knuckle joint press: A short-stroke press in which the slide is actuated by a crank attached to a knuckle joint hinge.

Knuckle post: A post which perform as the pivot for the steering knuckle in an automobile.

Kolosov-Muskhelishvili formulas: The Kolosov-Muskhelishvili formulas provide the most convenient method to treat two-dimensional crack problems. This formulation allows us to consider stresses and displacements in terms of analytic functions of complex variables.

Krigar-Menzel law: A generalization of the second Young-Helmholtz law which states that when a string is bowed at a point which is at a distance of p/q times the string's length from one of the ends, where p and q are relative primes, then the string moves back and forth with two constant velocities.

Kullenberg piston corer: The Kullenberg corer is a single-drive, wire-deployed piston corer that is dropped into the sediment from a short distance, propelled by the momentum of the heavy lead weights on the core head.

Laboratory coordinate system: A frame of reference attached to the laboratory of the observer, as opposed to the center of mass system.

Ladder: An arrangement, often portable, for climbing up and down; it consists of two parallel sides joined by a series of crosspieces that serve as footrests.

Ladder drilling: An arrangement of retractable drills with pneumatic powered legs mounted on banks of steel ladders connected to a holding frame; used in large-scale rock tunneling, with the advantage that many drills can be worked at the same time by a small labor force.

Ladder trencher: A machine that digs trenches by a bucket ladder excavator.

Lagrange bracket: Lagrange brackets are certain expressions closely related to Poisson brackets that were introduced by Joseph Louis Lagrange in 1808–1810 for the purposes of mathematical formulation of classical mechanics.

Lagrange-Hamilton theory: The formalized cram of continuous systems in terms of field variables where a Lagrangian density function and Hamiltonian density function are introduced to produce equations of motion.

Lagrange's equations: Equations of motion of a mechanical system for which a classical (non-quantum-mechanical) description is suitable, and which relate the kinetic energy of the system to the generalized coordinates, the generalized forces, and the time.

Lagrangian: The difference between the kinetic energy and the potential energy of a system of particles, expressed as a function of generalized coordinates and velocities from which Lagrange's equations can be imitative.

Lagrangian density: For a dynamical system of fields or continuous media, a function of the fields, of their time and space derivatives, and the coordinates and time, whose integral over space is the Lagrangian.

Lagrangian function: The function which measures the difference between the kinetic and potential energy of a dynamical arrangement.

Lami's theorem: When three forces act on a body in equilibrium, the magnitude of each is proportional to the sine of the angle between the other two.

Lancashire boiler: A cylindrical steam boiler having two longitudinal furnace tubes which have internal grates at the front.

lance door: The door to a boiler furnace through which a hand lance is inserted.

Lanchester balancer: A tool for balancing four-cylinder engines; consists of two meshed gears with eccentric masses, driven by the crankshaft.

Lanchester's rule: The rule that a torque applied to a rotating body along an axis perpendicular to the rotation axis will produce precession in a direction such that, if the body is viewed along a line of sight coincident with the torque axis, then a point on the body's circumference,

Landing gear: A pair of small wheels at the forward end of a semitrailer to carry the vehicle when it is detached from the tractor.

Land measure: Units of area used in measuring land. Any system for measuring land.

Latent heat: Latent heat, energy absorbed or re-leased by a substance during a change in its physical state (phase) that occurs without changing its temperature.

Latent load: Cooling required to remove unwanted moisture from an air conditioned space.

Lathe: A machine for shaping a work piece by gripping it in a holding device and rotating it under power against a suitable cutting tool for turning, boring, facing, or threading

Law of corresponding times: The principle that the times for corresponding motions of dynamically similar systems are proportional to L/V and also to (L/F), where L is a typical dimension of the system, V a typical velocity, and F a typical force per unit mass.

Leading truck: A swiveling frame with wheels under the front end of a locomotive.

Lead screw: A threaded shaft used to convert rotation to longitudinal motion, in a lathe it moves the tool carriage when cutting threads; in a disk recorder it guides the cutter at a desired rate across the surface of an un grooved disk.

League: A unit of length equal to 3 miles or 4828.032 meters.

Lean mixture: Fuel air mixture containing a low percentage of fuel and a high percentage of air, as compared with a normal or rich mixture.

Least energy principle: The principle that the potential energy of a system in stable equilibrium is a minimum relative to that of nearby configurations.

Least work theory: A theory of statically indeterminate structures based on the fact that when a stress is applied to such a structure the individual parts of it are deflected so that the energy stored in the elastic members is minimized.

Lee's disk: A device for determining the thermal conductivity of poor conductors in which a thin, cylindrical slice of the substance under study is sandwiched between two copper disks, a heating coil is placed between one of these disks and a third copper disk, and the temperatures of the three copper disks are measured.

Leidenfrost point: The smallest temperature at which a hot body submerged in a pool of boiling water is completely blanketed by a vapor film; there is a minimum in the heat flux from the body to the water at this temperature

Leidenfrost's phenomenon: A phenomenon in which a liquid dropped on a surface that is above a critical temperature becomes insulated from the surface by a layer of vapor, and does not wet the surface as a result.

Length: Length is a measure of distance. In the International System of Quantities, length is a quantity with dimension distance. In most systems of measurement a base unit for length is chosen, from which all other units are derived. In the International System of Units system the base unit for length is the metre.

Leslie cube: A metal box, with faces having different surface finishes, in which water is heated and next to which a thermopile is placed in order to compare the heat emission properties of different surfaces.

Level measurement: The determination of the linear vertical distance between a reference point or datum plane and the surface of a liquid or the top of a pile of divided solid.

Level valve: A valve worked by a lever which travels through a maximum arc of 180.

Leverage: The multiplication of force or motion attained by a lever.

Levitated vehicle: A train or other vehicle which travels at high speed at some distance above an electrically conducting track by means of levitation.

L-head engine: A type of four stroke cycle internal combustion engine having both inlet and exhaust valves on one side of the engine block which are operated by pushrods actuated by a single camshaft.

Lifting block: A combination of pulleys and ropes which allows heavy weights to be lifted with least effort.

Lift valve: A valve that moves perpendicularly to the plane of the valve seat laboratory coordinate system. A reference frame attached to the laboratory of the observer, in contrast to the center-of-mass system.

Light inspection car: A railway motorcar weighing 400-600 pounds (180-270 kilo-grams) and having a capacity of 650-800 pounds (295-360 kilograms).

light section car: A railway motorcar weighing 750-900 pounds (340-408 kilograms) and propelled by 4-6-horsepower (3000-4500-Watt) engines.

Lilly controller: An apparatus on steam and electric winding engines that protects against overspeed, overwind, and other incidents injurious to workers and the engine.

Limit control: In boiler operation, usually a device, electrically controlled, that shuts down a burner at a prescribed operating point.

Limited-rotation hydraulic actuator: A type of hydraulic actuator that produces limited reciprocating rotary force and motion; used for lifting, lowering, opening, closing, indexing, and transferring movements; examples are the piston-rack actuator, single-vane actuator, and double-vane actuator.

Limit governor: A mechanical governor that takes control of the chief governor to switch off the machine when the velocity reaches a predetermined excess above the allowable rate.

Limit velocity: In arm or

and projectile testing, the lowest possible velocity at which any one of the complete penetrations is obtained.

Linear actuator: A tool that converts some kind of power, such as hydraulic or electric power, into linear motion.

Linear strain: The ratio of the change in the length of a body to its original length. Also known as longitudinal strain.

Line of action: The locus of contact points as gear teeth profiles go through mesh.

Line of fall : The line tangent to the ballistic trajectory at the level point

Line of flight: The line of movement, or the intended line of movement, of an aircraft, guided missile, or projectile in the air

Line of impact: A line tangent to the trajectory of a missile at the point of impact.

Linkage: A device that transfers motion in a desired manner by using some combination of bar links, slides, pivots, and rotating members

Linter: A machine for removing fuzz linters from ginned cottonseed.

Liquid cooled engine: An internal combustion engine with a jacket cooling arrangement in which liquid, usually water, is circulated to maintain acceptable operating temperatures of machine parts.

Liquid measure: A system of units used to measure the volumes of liquid substances in the United States; the units are the fluid dram, fluid ounce, gill, pint, quart, and gallon

Liquid piston rotary compressor: A rotary compressor in which a multi blade rotor revolves in a casing partly filled with liquid, for example, water

Liquid sorbent dehumidifier: A sorbent type of dehumidifier consisting of a main circulating fan, sorbent-air contactor, sorbent pump, and reactivator; dehumidification and reactivation are continuous operations, with a small part of the sorbent constantly bled off from the main circulating system and reactivated to the concentration required for the desired effluent dew point.

Liquidus line: For a two-component system, a curve on a graph of temperature versus concentration which connects temperatures at which fusion is completed as the temperature is raised

Liter: The unit of volume or capacity, equal to 1 decimeter cubed, or 0.001 cubic meter, or 1000 cubic centimetres.

Live axle: An axle to which wheels are rigidly fixed.

Live center: A lathe center that fits into the headstock spindle.

Live load: A moving load or a load of variable force acting upon a structure, in addition to its own weight.

Live roller conveyor: Conveying machine which budge objects over a series of rollers by the application of power to all or some of the rollers.

Live steam: Steam that is being distributed directly from a boiler under full pressure.

Livre: A unit of mass, used in France, equal to 0.5 kilogram.

Load and carry equipment: Earthmoving tools designed to load and transport material.

Load carrying capacity: The maximum weight that the end effector of a robot can manipulate without reducing its level of performance

Load deflection: The change in position of a body when a load is applied to it.

Loader: A machine such as a mechanical shovel used for loading bulk materials

Loading head: The part of a loader which gathers the bulk materials.

Loading station: A device which receives material and puts it on a conveyor. It may be one or more plates or a hopper

Load stress: Stress that grades from a pressure or gravitational load.

Local buckling : Buckling of thin elements of a column section in a series of waves or wrinkles

local coefficient of heat transfer: Heat-transfer coefficient at a particular point on the heat-transfer surface, equal to the local heat flux at this point (q_w) divided by the local temperature drop (Δt).

Localized vector: A vector whose line of application or point of application is prescribed, in addition to its direction.

Locating: A function of tooling action accomplished by designing and constructing the tooling device so as to bring together the proper contact points or surfaces between the work piece and the tooling

Locating hole: A hole used to position the part in relation to a cutting tool or to other parts and gage points.

Locating surface: A surface inured to position an item being manufactured in a numerical control or robotic system for clamping.

Locomotive: A self-propelling machine with flanged wheels, for moving loads on railroad tracks; utilizes fuel (for steam or internal combustion engines), compressed air, or electric energy

Locomotive boiler: An internally fixed horizontal fire-tube boiler with integral furnace; the doubled furnace walls contain water which mixes with water in the boiler shell

Locomotive crane: A crane mounted on a railroad flatcar or a special chassis with flanged wheels.

Longitudinal acceleration: The component of the linear acceleration of an aircraft, missile, or particle parallel to its longitudinal, or X, axis.

Longitudinal drum boil-

er: A boiler in which the axis of the horizontal drum is parallel to the tubes, both lying in the same plane.

Longitudinal vibration: A continuing periodic change in the displacement of parts of a rod-shaped object in the direction of the long axis of the rod.

Loose pulley: In belt-driven machinery, a pulley which turns freely on a shaft so that the belt can be shifted from the driving pulley to the loose pulley, thereby causing the machine to stop.

Loss-in-weight feeder: A device to apportion the output of granulated or powdered solids at a constant rate from a feed hopper; weight measured decrease in hopper content actuates further opening of the discharge chute to compensate for flow loss as the hopper overburden decreases; used in the chemical, fertilizer, and plastics industries.

Lost motion: The delay between the movement of a driver and the movement of a follower.

Lowboy: A trailer with low ground clearance for hauling construction equipment.

Lower pair: A link in a

mechanism in which the mating parts have surface (instead of line or point) contact.

Lowest safe water line: The lowest water level in a boiler drum at which the burner may safely work.

Low heat value: The heat value of a combustion process assuming that none of the water vapor resulting from the process is condensed out, so that its latent heat is not available.

Low intensity atomizer: A type of electrostatic atomizer operating on the principle that atomization is the result of Rayleigh instability, in which the presence of charge in the surface counteracts surface tension

Low-level condenser: A direct contact water cooled steam condenser that uses a pump to remove liquid from a vacuum space.

Low lift truck: A hand or powered lift truck that lift the load sufficiently to make it mobile.

Low pressure area: The point in a bearing where the pressure is the least and the area or space for a lubricant is the greatest.

Low water fuel cutoff: A float apparatus which shuts off fuel supply and burner when boiler water level drops below the lowest safe waterline.

Ludwig Soret effect: An occurrence in which a temperature gradient in a mixture of substances gives rise to a concentration gradient.

Machine: An arrangement of rigid or resistant bodies with definite motions and capable of performing useful work.

Machine drill: A drill or drilling machine is a tool primarily used for making round holes or driving fasteners. Any mechanically driven diamond, rotary, or percussive drill.

Machine rating: The power that a machine can draw or convey without overheating.

Machinery: A group of parts or machines arranged to perform a useful function.

Machine shop: A workshop where work, metal or other material is machined to specified size and assembled.

Machine taper: A taper that provides a connection between a tool, arbor, or center and its mating part to ensure and maintain accurate alignment between the parts; permits easy separation of parts.

Machine tool: Stationary power-driven machine for forming, cutting, turning, boring, drilling, grinding or polishing hard parts, especially metals.

Machining: Performing different cutting or grinding operations on a piece of work.

Machining center: Manufacturing tools that removes metal under computer numerical control by making use of several axes and a variety of tools and operations

Magnetic bearing: A device incorporating magnetic forces to cause a shaft to levitate and float in a magnetic field without any contact between the rotating and stationary elements.

Magnetic brake: A friction brake under the control of an electromagnet.

Magnetic chuck: A chuck in which the workpiece is held by magnetic force.

Magnetic fluid clutch: A friction clutch that is engaged by magnetizing a liquid suspension of powdered iron located between pole pieces mounted on the input and output shafts.

Magnetic friction clutch: A friction clutch in which the pressure between the friction surfaces is produced by magnetic lure. Also known as magnetic clutch.

Magnetocaloric effect: The reversible modify of temperature accompanying the change of magnetization of a ferromagnetic material.

Main bearing: One of the bearings that carry the crankshaft in an internal combustion engine

Main shaft: The line of shafting receiving its power from the engine or motor and transmitting power to other parts.

Mandrel press: A press for driving mandrels into holes.

Mangle gearing: Gearing for creating reciprocating motion; a pinion rotating in a single direction drives a rack with teeth at the ends and on both sides

Manocryometer: An instrument for measuring the change of a substance's melting point with change in pressure.

Manometer: A manometer is a device to measure pressures. A common simple manometer consists of a U shaped tube of glass filled with some liquid.

Manometry: The use of manometers to measure gas and vapor pressures.

Mass: Quantitative measure of inertia, a fundamental property of all matter. It is, in effect, the resistance that a body of matter offers to a change in its speed or position upon the application of a force. The greater the mass of a body, the smaller the change produced by an applied force.

Massieu function: The negative of the Helmholtz free energy divided by the temperature.

Mass units: Units of measurement having to do with masses of materials, such as pounds or grams.

Master cylinder: The container for the fluid and the piston, forming part of a device such as a hydraulic brake or clutch.

Material particle: A particle is a small localized object to which can be ascribed several physical or chemical properties, such as volume, density, or mass.

Matthiessen sinker method: Process of determining the thermal expansion coefficient of a liquid, in which the apparent weight of a sinker when immersed in the liquid is measured for two different temperatures of the liquid.

Maupertius' principle: The principle of least action is sufficient to conclude the motion of a mechanical system.

Maximum allowable working pressure: The utmost gage pressure in a pressure vessel at a designated temperature, used for the determination of the set pressure for relief valves.

Maximum angle of inclination: The highest angle at which a conveyor may be inclined and still deliver an amount of bulk material within a given time.

Maximum belt slope: A slope beyond which the material on the belt of a conveyor tends to roll downhill.

Maximum belt tension: The total of the starting and operating tensions. In the average conveyor this is considered to be the same as the tight side tension.

Maximum continuous load: The greatest load that a boiler can maintain for a designated length of time.

Maximum gradability: Steepest slope a vehicle can negotiate in low gear; usually expressed in percentage of slope, namely, the ratio between the vertical rise and the horizontal distance traveled; sometimes expressed by the angle between the slope and the horizontal

Maximum ordinate: Distinction in altitude between the origin and highest point of the trajectory of a projectile

Maximum production life: The length of time that a cutting tool execute at cutting conditions of maximum tool efficiency.

Maxwell equal area rule: At temperatures for which the theoretical isothermal of a substance, on a graph of pressure against volume.

Maxwell's theorem: If a load applied at one point A of an elastic structure results in a given deflection at another point B, then the same load applied at B will result in the same deflection at A.

Mayer: A unit of heat capacity equal to the heat capacity of a substance whose temperature is raised 1 Celsius by 1 joule.

Mean calorie: One hundredth of the heat needed to raise 1 gram of water from 0 to 100 C.

Mean effective pressure: A term generally used in the evaluation for positive displacement machinery performance which expresses the average net pressure difference in pounds per square inch on the two sides of the piston in engines, pumps, and compressors.

Mean normal stress: In a system stressed multi axially, the algebraic mean of the three principal stresses.

Mean specific heat: The average over a specified range of temperature of the specific heat of a matter.

Mean stress: The algebraic mean of the maximum and minimum values of a periodically varying stress.

Mean trajectory: The trajectory of a missile that passes through the center of impact or center of burst.

Mechanical advantage: The ratio of weight lifted by a machine such as a lever or pulley to the force applied to it.

Mechanical analysis: A sorting operation in which mixtures of particles of mixed sizes, and often of different specific gravities, are separated into fractions by the action of a stream of fluid, usually water.

Mechanical classifier: Machines that are commonly used to classify mixtures of particles of different sizes, and sometimes of different specific gravities.

Mechanical draft: A draft that depends upon the use of fans or other mechanical devices; may be induced or forced.

Mechanical draft cooling tower: Mechanical draft cooling towers are the most widely used in buildings and rely on power-driven fans to draw or force the air through the tower. They are normally located outside the building. The two most common types of mechanical draft towers to the HVAC industry are induced draft and forced draft.

Mechanical efficiency: In an engine, the ratio of brake horsepower to indicated horsepower.

Mechanical Engineering: Mechanical engineering is an engineering branch that combines engineering physics and mathematics principles with materials science to design, analyze, manufacture, and maintain mechanical systems.

Mechanical equivalent of heat: The amount of mechanical energy equivalent to a unit of heat.

Mechanical gripper: A robot component that uses movable, finger like levers to grasp objects.

Mechanical hysteresis:

The dependence of the strain of a material not only on the instantaneous value of the stress but also on the previous history of the stress; for example, the elongation is less at a given value of tension when the tension is increasing than when it is decreasing.

Mechanical impedance: The complex ratio of a phasor representing a sinusoidally varying force applied to a system to a phasor representing the velocity of a point in the system.

Mechanical linkage: A set of rigid elements, called links, joined together at pivots by means of pins or equivalent devices.

Mechanical loader: A power machine used for loading mineral, coal, or dirt.

Mechanical ohm: A unit of mechanical resistance, reactance, and impedance, equal to a force of 1 dyne divided by a velocity of 1 centimeter per second.

Mechanical press: A press whose slide is activated by mechanical means.

Mechanical property: A property that involves a relationship between stress and strain or a reaction to an applied force.

Mechanical pulping: Mechanical pulping is the process in which wood is separated or defibrated mechanically into pulp for the paper industry. The mechanical pulping processes use wood in the form of logs or chips that are mechanically processes.

Mechanical pump: A pump through which fluid is conveyed by direct contact with a moving part of the pumping machinery.

Mechanical reactance: The imaginary part of mechanical impedance.

Mechanical refrigeration: The removal of heat by utilizing a system as refrigerator and medium as refrigerant, subjected to cycles and employing a mechanical compressor.

Mechanical seal: Mechanical assembly that forms a leak proof seal between flat, rotating surfaces to prevent high-pressure leakage.

Mechanical separation: A group of industrial operations by means of which particles of solid or drops of liquid are removed from a gas or liquid, or are separated into individual fractions, or both, by gravity separation (settling), centrifugal action, and filtration

Mechanical setting: Producing bits by setting diamonds in a bit mold into which a cast or powder metal is placed, thus embedding the diamonds and forming the bit crown; opposed to hand setting

Mechanical shovel: A loader limited to level or slightly graded drivages; when full, the shovel is swung over the machine, and the load is discharged into containers or vehicles behind.

Mechanical torque converter: A torque converter, such as a pair of gears, that convey power with only incidental losses

Mechanical units: Units of length, time, and mass, and of physical quantities derivable from them.

Mechanical vibration: The continuing motion, often repetitive and periodic, of parts of machines and structures.

Mechanism: That part of a machine which have two or more pieces so arranged that the motion of one compels the motion of the others.

Mechanize: To substitute machinery for human or animal labor.

Mechanomotive force: The root mean square value of a periodically varying force.

Mega second: A unit of time, equal to 1,000,000 seconds.

Megawatt: A unit of power, equal to 1,000,000 Watts.

Melde's experiment: Melde's experiment is a scientific experiment carried out in 1859 by the German physicist Franz Melde on the standing waves produced in a tense cable originally set oscillating by a tuning fork, later improved with connection to an electric vibrator.

Melt fracture: Melt flow instability through a die during plastics molding, leading to helicular, rippled surface irregularities on the finished product.

Melting furnace: A furnace in which the frit for glass is melted.

Melting point: The temperature at which a solid of a pure substance changes to a liquid.

Melt instability: Instability of the plastic melt flow through a die.

Melt strength: Strength of

a molten plastic.

Membrane analogy: Formal identity between the differential equation and boundary conditions for the stress function for torsion of an elastic prismatic rod and the conditions for deflection of a uniformly stretched membrane with the same boundary as the cross-section of the rod under the action of uniform pressure.

Membrane stress: Stress which is equivalent to the average stress across the cross section involved and normal to the reference plane.

Mercer engine: A revolving block engine in which two opposing pistons work in a single cylinder with two rollers attached to each piston; intake ports are uncovered when the pistons are closest together, and exhaust ports are uncovered when they are farthest apart

Metal-slitting saw: A milling cutter similar to a circular saw blade but sometimes with side teeth as well as teeth around the circumference; used for deep slotting and sinking in cuts.

Metarheology: A branch of rheology whose approach is intermediate between those of macrorheology and mi-

crorheology.

Meter: The international standard unit of length.

Meter bar: A metal bar for mounting a gas meter, having fittings at the ends for the inlet and outlet connections of the meter.

Metering screw: An extrusion type screw feeder or conveyor section used to feed pulverized or doughy material at a constant rate.

Metering valve: In an automotive hydraulic braking system, a valve that momentarily delays application of the front disk brakes until the rear drum brakes begin to act.

Meter-kilogram-second system: A metric system of units in which length, mass, and time are fundamental quantities, and the units of these quantities are the meter, the kilogram, and the second respectively.

Meter stop: A valve installed in a water service pipe for control of the flow of water to a building.

Method of mixtures: A method of determining the heat of fusion of a substance whose specific heat is known, in which a known amount of the solid is combined with a known amount of the liquid in a calorimeter, and the decrease in the liquid temperature during melting of the solid is measured.

Metric centner: A unit of mass equal to 50 kilograms.

Metric grain: A unit of mass, equal to 50 milligrams; used in commercial transactions in precious stones.

Metric system: A system of units used in scientific work throughout the world and employed in general commercial transactions and engineering applications.

Metric-technical unit of mass: A unit of mass equal to mass that is accelerated by 1 meter per second per second by a force of 1 kilogram-force; it is equal to 9.80665 kilograms. Abbreviated as TME.

Microangstrom: A unit of length equal to one-millionth of an angstrom, or 10^{-16} meter.

Microgram: A unit of mass equal to one millionth of a gram.

Microgravity: A state of very weak gravity, in which the gravitational acceleration experienced by the observer inside the system under consideration is on the order of one millionth of the Earth's.

Micron: A unit of pressure equal to the pressure exerted by a column of mercury 1 micrometer high,

Microrheology: A branch of rheology in which the heterogeneous nature of dispersed systems is taken into account.

Micro second: A unit of time equal to one millionth of a second

Micro watt: A unit of power equal to one millionth of a watt.

Mile: A unit of length in common use in the United States, equal to 5280 feet, or 1609.344 meters.

Milligram: A unit of mass equal to one thousandth of a gram.

Milliliter: A unit of volume equal to 10^{-3} liter or 10^{-6} cubic meter. Abbreviated ml.

Millimeter: A unit of length equal to one thousandth of a meter. Abbreviated mm.

Millimeter of mercury: A millimetre of mercury is a manometric unit of pressure, formerly defined as the extra pressure generated by a column of mercury one millimetre high, and currently defined as exactly

133.322387415 pascals.

Millimeter of water: A unit of pressure, equal to the pressure exerted by a column of water 1 millimeter high with a density of 1 gram per cubic centimeter under the standard acceleration of gravity.

Milling: Milling is the process of machining using rotary cutters to remove material by advancing a cutter into a workpiece. This may be done varying direction on one or several axes, cutter head speed, and pressure.

Milling machine: A machine for the removal of metal by feeding a work piece through the periphery of a rotating circular cutter. Also known as miller.

Milling planer: A planer that employ a rotary cutter rather than single-point tools.

Milli second: A unit of time equal to one thousandth of a second.

Milliwatt: A unit of power equal to one thousandth of a watt. Abbreviated mW.

Mine car: An industrial car, usually of the four-wheel type, with a low body; the door is at one end

Minute: A unit of time, equal to 60 seconds.

Mired: A unit used to measure the reciprocal of color temperature, equal to the reciprocal of a color temperature of 10^6 kelvins. Derived from micro-reciprocal-degree.

Missile attitude: The location of a missile as determined by the inclination of its axes (roll, pitch, and yaw) in relation to another object, as to the earth

Mistuning: The disparity between the square of the natural frequency of vibration of a vibrating system, without the effect of damping, and the square of the frequency of an external, oscillating force

Mixed cycle: An internal combustion engine cycle which combines the Otto cycle constant-volume combustion and the Diesel cycle constant-pressure combustion in high-speed compression-ignition engines. Also known as combination cycle; commercial Diesel cycle; limited-pressure cycle

Mixed-flow impeller: An impeller for a pump or compressor which combines radial and axial flow principles.

M meter: A class of instruments which measure the

liquid water content of the atmosphere.

Mobile crane: A cable controlled crane mounted on crawlers or rubber tired carriers.

Mobile hoist: A platform hoist mounted on a pair of pneumatic tired road wheels, so it can be towed from one site to another.

Mobile loader: A self-propelling power machine for loading coal, mineral, or dirt.

Mode of vibration: A typical way, in which a system that does not dissipate energy and whose movements are limited by boundary conditions, can oscillate, having a characteristic pattern of movement and one of a discrete set of frequencies. Also known as oscillation mode.

Modulation: Regulation of the fuel air mixture to a burner in response to fluctuations of load on a boiler.

Modulus of decay: The time required for the amplitude of the undamped harmonic oscillator to fall to 1 / e from its initial value; reciprocal of the damping factor.

Modulus of deformation: The modulus of elasticity of a material that deforms other than according to Hooke's law.

Modulus of elasticity: The ratio of the increment of some specified form of stress to the increment of some specified form of strain, such as Young's modulus, the bulk modulus, or the shear modulus

Modulus of elasticity in shear: A measure of a material's resistance to shearing stress, equal to the shearing stress divided by the resultant angle of deformation expressed in radians

Modulus of resilience: The maximum mechanical energy stored per unit volume of material when it is stressed to its elastic limit.

Modulus of rupture in bending: The maximum stress per unit area that a specimen can withstand without breaking when it is bent, as calculated from the breaking load under the assumption that the specimen is elastic until rupture takes place

Modulus of rupture in torsion: The maximum stress per unit area that a specimen can withstand without breaking when its ends are twisted, as calculated from the breaking load under the assumption that the specimen is elastic until rupture takes place

Mohm: A unit of mechanical mobility, equal to the reciprocal of 1 mechanical ohm.

Mohr's circle: A graphical construction making it possible to determine the stresses in a cross section if the principal stresses are known.

Moisture content: The amount of water in a mass of soil, sewage, sludge, or screenings; expressed in percentage by weight of water in the mass.

Moisture loss: The disparity in heat content between the moisture in the boiler exit gases and that of moisture at ambient air temperature

Molecular heat: The heat capacity per mole of a substance.

Molecular heat diffusion: Transfer of heat through the motion of molecules.

Molecular pump: A vacuum pump in which the molecules of the gas to be exhausted are carried away by the friction between them and a rapidly revolving disk or drum.

Mollier diagram: Graph of enthalpy versus entropy of a vapor on which isobars, isothermals, and dryness fraction lines are there.

Moment: Static moment of some quantity, except in the term "moment of inertia

Momental ellipsoid: An ellipsoid of inertia, the size of which is set such that the vertex of the angular velocity vector of a freely rotating object with the origin at the center of the ellipsoid always lies on the surface of the ellipsoid. Also known as energy ellipsoid.

Moment diagram: A diagram of the bending moment at a section of a beam versus the distance of the section along the beam.

Moment of inertia: The moment of inertia, otherwise known as the mass moment of inertia, angular mass, second moment of mass, or most accurately, rotational inertia, of a rigid body is a quantity that determines the torque needed for a desired angular acceleration about a rotational axis.

Momentum: Momentum can be defined as "mass in motion." All objects have mass; so if an object is moving, then it has momentum - it has its mass in motion. The amount of momentum that an object has is dependent upon two variables: how much *stuff* is moving and how fast the *stuff* is moving.

Monocable: An aerial rope-way that uses one rope to both support and haul a load.

Monochromatic emissivity: The ratio of the energy radiated by a body in a very narrow band of wavelengths to the energy radiated by a blackbody in the same band at the same temperature.

Monochromatic temperature scale: A temperature scale based upon the amount of power radiated from a blackbody at a single wavelength

Moody formula: A formula giving the efficiency of a field turbine, whose runner has diameter D , in terms of the efficiency of a model turbine

Morgan equation: A modification of the Ramsey-Shields equation, in which the expression for the molar surface energy is set equal to a quadratic function of the temperature rather than to a linear one.

Mortising machine: A machine that uses a screw and chisel to make a square or rectangular groove in wood.

Motion: A continuous change of position of a body.

Motorcycle: An automotive vehicle, essentially a motorized

bicycle, with two tandem and sometimes three rubber wheels.

Motor reducer: Equipment for power transmission with reduced speed, in which gearboxes are an integral part of the drive motors.

Motor truck: An automotive vehicle which is used to transport freight.

Motor vehicle: Any automotive vehicle that does not run on rails, and generally having rubber tires.

Mounce: A unit of mass, equal to 25 grams.

Movable-active tooling: Any tools in a robotic system that is able to move and that operates under power

Movable-passive tooling: Equipment in a robotic arrangement that moves but requires no power to operate, such as workpieces, clamps, and templates.

Moving constraint: A constraint that changes with time, as in the case of a system on a moving platform.

Moving load: A load that can move, such as vehicles or pedestrians.

Multiple mid stop: A peripheral device that allows a pick-and-place robot to swing and stop in several positions.

Multiple-slide press: Press with individual adjustable slide, built into the main slide or independently connected to the main shaft.

Multiple-strand conveyor: A conveyor with two or more spaced strands of chain, belts, or cords as the supporting or propelling medium.

Multirope friction winder: A winding system in which the drive to the winding ropes is the frictional resistance between the ropes and the driving sheaves.

Multistage compressor: A machine for compressing a gaseous fluid in a sequence of stages, with inter cooling between stages.

Multistage pump: A pump in which the head is developed by multiple impellers operating in series.

Muskhelishvili's method: A method for solving problems of elastic deformation of a plane body, including the use of methods of the theory of functions of a complex variable for calculating analytical functions that determine the plane deformation of a body.

N

Nano gram: A measure of weight. One billionth (10^{-9}) of a gram.

Nano meter: A unit of length equal to one-billionth of a meter,

Nano second: A unit of time equal to one billionth of a second, or 10^{-9} second.

Nano technology: Branch of Science that deals with transforming matter, energy, and information that are based on nanometer scale components with precisely defined molecular features. Techniques that produce or measure features less than 100 nanometers in size.

Natural convection: Natural convection is a type of flow, of motion of a liquid such as water or a gas such as air, in which the fluid motion is not generated by any external source but by density difference.

Natural draft cooling tower: A cooling tower that depends upon natural convection of air flowing upward and in contact with the water to be cooled.

Natural gasoline plant: Compression, distillation, and absorption process facility used to remove natural gasoline (mostly butanes and heavier components) from natural gas.

Nautical chain: Unit of length equal to 15 feet or 4.572 meters.

Navier's equation: Navier equations are used to describe the deformation of a homogeneous, isotropic and linear elastic medium in the absence of body forces.

Needle nozzle: A streamlined hydraulic turbine nozzle with a movable element for converting the pressure and kinetic energy in the pipe leading from the reservoir to the turbine into a smooth jet of variable diameter and discharge but practically constant velocity.

Needle valve: A slender, pointed rod fitting in a hole or circular or conoidal seat.

Negative rake: The orientation of a cutting tool whose cutting edge lags the surface of the tooth face.

Negative temperature: The property of a thermally isolated thermodynamic system whose elements are in thermodynamic equilibrium among themselves.

Nernst approximation formula: An equation for

the equilibrium constant of a gas reaction derived from the Nernst heat theorem and certain simplifying assumptions.

Nernst heat theorem: The theorem expressing that the rate of change of free energy of a homogeneous system with temperature, and also the rate of change of enthalpy with temperature, approaches zero as the temperature approaches absolute zero.

Nernst-Simon statement of the third law of thermodynamics: The statement that the change in entropy which occurs when a homogeneous system undergoes an isothermal reversible process approaches zero as the temperature approaches absolute zero.

Net positive suction head: The minimum suction head required for a pump to operate, it depends on liquid characteristics, total liquid head, pump speed and capacity, and impeller design.

Neumann Kopp rule: The rule that the heat capacity of 1 mole of a solid substance is approximately equal to the sum over the elements forming the substance of the heat capacity of a gram atom of the element times the number of atoms of the element in a molecule of the substance.

Neutral axis: In a beam bent downward, the line of zero stress below which all fibers are in tension and above which they are in compression.

Neutral surface: A surface in a bent beam along which material is neither compressed nor extended.

Newton: The unit of force in the meterkilogram-second system, equal to the force which will impart an acceleration of 1 meter per second squared to the International Prototype Kilogram mass. Symbolized N.

Newtonian attraction: The common attraction of any two particles in the universe, as given by Newton's law of gravitation.

Newtonian mechanics: The system of mechanics based upon Newton's laws of motion in which mass and energy are considered as separate, conservative, mechanical properties, in contrast to their treatment in relativistic mechanics.

Newtonian reference frame: One of a set of reference frames with constant relative velocity and within which Newton's laws hold; the frames have a common time, and coordinates are related by the Galilean transformation rule.

Newtonian velocity: Relative velocity is a measurement of velocity between two objects as determined in a single coordinate system.

Newton-meter of torque: The unit of torque in the meter-kilogram-second system, equal to the torque produced by 1 newton of force acting at a perpendicular distance of 1 meter from an axis of rotation.

Newton's equations of motion: Newton's laws of motion expressed in the form of mathematical equations.

Newton's first law: The law that a particle not subjected to external forces remains at rest or moves with constant speed in a straight line.

Newton's law of cooling: The law that the rate of heat flow out of an object by both natural convection and radiation is proportional to the temperature difference between the object and its environment, and to the surface area of the object.

Newton's law of gravitation: The law that every two

particles of matter in the universe attract each other with a force that acts along the line joining them, and has a magnitude proportional to the product of their masses and inversely proportional to the square of the distance between them.

Newton's laws of motion: Three fundamental principles (called Newton's first, second, and third laws) which form the basis of classical, or Newtonian, mechanics, and have proved valid for all mechanical problems not involving speeds comparable with the speed of light and not involving atomic or subatomic particles.

Newton's second law: The law that the acceleration of a particle is directly proportional to the resultant external force acting on the particle and is inversely proportional to the mass of the particle. Also known as second law of motion.

Newton's third law: The law that, if two particles interact, the force exerted by the first particle on the second particle (called the action force) is equal in magnitude and opposite in direction to the force exerted by the second particle on the first particle.

Nibbling: Contour cutting of

material by the action of a reciprocating punch that takes repeated small bites as the work is passed beneath it.

Noise equivalent temperature difference: The change in equivalent blackbody temperature that corresponds to a change in radiance which will produce a signal-to-noise ratio of 1 in an infrared imaging device.

Non-black body: A body that reflects some fraction of the radiation incident upon it; all real bodies are of this nature.

Nonequilibrium thermodynamics: A quantitative treatment of irreversible processes and of rates at which they occur. Also known as irreversible thermodynamics.

Non holonomic system: A system of particles which is subjected to constraints of such a nature that the system cannot be described by independent coordinates; examples are a rolling hoop, or an ice skate which must point along its path.

Non integrable system: A dynamical arrangement whose motion is governed by an equation that is not an integrable differential equation.

Nonlinear vibration: A vibration whose amplitude is maximum so that the elastic restoring force on the vibrating object is not proportional to its displacement

Non quantum mechanics: The classical mechanics of Newton and Einstein as opposed to the quantum mechanics of Heisenberg, Schrodinger, and Dirac; particles have definite position and velocity, and they move according to Newton's laws.

Non-reclosing pressure relief device: A device which remains open after relieving pressure and must be reset before it can operate again.

Non relativistic kinematics: The study of motions of systems of objects at speeds which are small compared to the speed of light, without reference to the forces which act on the system

Non relativistic mechanics: The study of the dynamics of arrangement in which all speeds are small compared to the speed of light.

Normal acceleration: The part of the linear acceleration of an aircraft or missile along its normal, or Z, axis

Normal axis: The vertical

axis of an aircraft or missile

Normal coordinates: A set of coordinates for a coupled system such that the equations of motion each involve only one of these coordinates.

Normal frequencies: The frequencies of the normal modes of vibration of a system.

Normal impact: Impact on a plane perpendicular to the trajectory.

Normal mode of vibration: Vibration of a coupled system in which the value of one of the normal coordinates oscillates and the values of all the other coordinates remain stationary.

Normal operation: The operation of a boiler or pressure vessel at or below the conditions of coincident pressure and temperature for which the vessel has been intended.

Normal pitch: The distance between working faces of two adjacent gear teeth, measured between the intersections of the line of action with the faces.

Normal reaction: The force exerted by a surface on an object in contact with it which prevents the object from passing through the surface; the force is perpendicular to the surface, and is the only force that the surface exerts on the object in the absence of frictional forces.

Normal stress: The stress component at a point in a structure which is perpendicular to the reference plane.

Nose radius: The radius measured in the back rake or top rake plane of a cutting tool.

Notching press: A mechanical press for notching straight or rounded edges.

Nozzle efficiency: The efficiency with which a nozzle converts potential energy into kinetic energy, commonly expressed as the ratio of the actual change in kinetic energy to the ideal change at the given pressure ratio.

Nuclear power plant: A power plant in which nuclear energy is transformed into heat for use in producing steam for turbines, which in turn drive generators that produce electric power.

Nusselt equation: Dimensionless equation used to calculate convection heat transfer for heating or cooling of fluids.

Nusselt number: A dimensionless number used in

the study of forced convection. In fluid dynamics, the Nusselt number is the ratio of convective to conductive heat transfer at a boundary in a fluid. Symbolized as N_{Nu}.

Nutation: A bobbing or nodding up and down movement of a spinning rigid body, such as a top, as it processes about its vertical axis.

Oblique valve: A type of globe valve which are suitable for installation on wet risers in a building for firefighting purposes with permanently charged water from a pressurised supply.

Ocean thermal-energy conversion: it is Abbreviated by "OTEC". The conversion of energy arising from the temperature difference between warm surface water of oceans and cold deep-ocean current into electrical energy or other useful forms of energy.

Octahedral normal stress: The normal component of stress across the faces of a regular octahedron whose vertices lie on the principal axes of stress. It is also known as mean stress.

Octahedral shear stress: The tangential component of stress across the faces of a regular octahedron whose vertices lie on the principal axes of stress; it is a measure of the strength of the deviatoric stress.

Octane requirement: The fuel octane number essential for efficient operation (without knocking or spark retardation) of an internal combustion engine.

Offhand grinding: Grinding operations performed with hand-held tools. It is also known as free hand grinding.

Off-highway vehicle: A bulk-handling machine, such as an earthmover or dump truck, that is intended to operate on steep or rough terrain and has a height and width that may exceed highway legal limits.

Off-road vehicle: A conveyance intended to travel on unpaved roads, trails, beaches, or rough terrain rather than on public roads.

Offset cylinder: A reciprocating part in which the crank rotates about a center off the center line.

Offset yield strength: Offset yield strength is an arbitrary approximation of a material's elastic limit. It is the stress that corresponds to a point at the intersection of a stress-strain curve and a line which is parallel to a specified modulus of elasticity line.

Oil cooler: A small radiator used to cool the oil that lubricates an automotive engine.

Oil dilution valve: A valve used to combine gasoline with engine oil to permit easier starting of the gasoline engine in cold weather.

Oil furnace: A combustion chamber in which oil is used as heat-producing fuel.

Oilless bearing: A self-lubricating bearing having solid or liquid lubricants in its material.

Oil lift: Hydrostatic lubrication of a journal bearing by using oil at high pressure in the area between the bottom of the journal and the bearing itself so that the shaft is raised and supported by an oil film whether it is rotating or not.

Oil pump: A pump of the gear, vane, or plunger type, usually an integral part of the automotive engine; it lifts oil from the sump to the upper level in the splash and circulating systems, and in forced-feed lubrication it pumps the oil to the tubes leading to the bearings and other parts.

Oil ring: A ring placed at the lower part of a piston to prevent an excess amount of oil from being drawn up onto the piston during the suction stroke.

Oleo strut: A shock absorber consisting of a telescoping cylinder that forces oil into an air chamber, thereby compressing the air; used on aircraft landing gear.

Once-through boiler: A boiler in which water flows, without recirculation, sequentially through the economizer, furnace wall, and evaporating and superheating tubes.

Onsager reciprocal relations: In thermodynamics, the Onsager reciprocal relations express the equality of certain ratios between flows and forces in thermodynamic systems out of equilibrium, but where a notion of local equilibrium exists.

Open-circuit grinding: Grinding system in which material passes through the grinder without classification of product and without recycle of oversize lumps; in contrast to closed-circuit grinding.

Open cycle: A thermodynamic cycle in which new mass enters the boundaries of the system and spent exhaust leaves it; the automotive engine and the gas turbine illustrate this process.

Open-cycle engine: An engine in which the working fluid is discharged after one pass through boiler and engine.

Open-cycle gas turbine: A gas turbine prime mover in which air is compressed in the compressor element, fuel is injected and burned in the combustor, and the hot products are expanded in the turbine element and exhausted to the atmosphere.

Opening die: A die head for cutting screws that opens automatically to release the cut thread.

Opening pressure: The static inlet pressure at which discharge is initiated.

Open system: A system in which both matter and energy may pass across boundaries.

Operating stress: The stress to which a structural unit is subjected in service.

Operating water level: The water level in a boiler drum which is usually maintained above the lowest safe level.

Opposed engine: An opposed-piston engine is a piston engine in which each cylinder has a piston at both ends, and no cylinder head.

Orbital angular momentum: The angular momentum associated with the motion of a particle about an origin, equal to the cross product of the position vector with the linear momentum.

Orbital plane: The plane which contains the orbit of a body or particle in a central force field; it passes through the center of force.

Orbital sander: An electric sander that travel the abrasive in an elliptical pattern.

Ordinary gear train: A gear train in which all axes remain stationary relative to the frame.

Orientation vector: A vector whose direction indicates the orientation of a robot gripper.

Orifice mixer: An Arrangement in which two or more liquids are pumped through an orifice constriction to cause turbulence and consequent mixing action.

Orthotropic: Having elastic possessions such as those of timber, that is, with considerable variations of strength in two or more directions perpendicular to one another.

Oscillating conveyor: A conveyor on which pulverized solids are moved by a pan or trough bed attached to a vibrator or oscillating method.

Oscillating granulator: Solids size reducer in which particles are broken by a set of oscillating bars arranged in cylindrical form over a screen of suitable mesh.

Oscillating screen: Solids separator in which the sifting screen oscillates at 300 to 400 revolutions per minute in a plane parallel to the screen.

Ostwald's adsorption isotherm: An equation stating that at a constant temperature the weight of material adsorbed on an adsorbent dispersed through a gas or solution, per unit weight of adsorbent, is proportional to the concentration of the adsorbent raised to some constant power.

Otto cycle: A thermodynamic cycle for the conversion of heat into work, heat addition takes place at constant volume process.

Otto engine: An internal combustion engine that operates on the Otto cycle, where the phases of suction, compression, combustion, expansion, and exhaust occur sequentially in a four-stroke-cycle or two-stroke-cycle reciprocating mechanism.

Otto-Lardillon method: A method of computing trajectories of missiles with low velocities (so that drag is proportional to the velocity squared) and quadrant angles of departure that may be high, in which exact solutions of the equations of motion are arrived at by numerical integration and are then tabulated.

Ounce: A unit of mass in avoirdupois measure equal to 1/16 pound or to approximately 0.0283495 kilogram.

Ouncedal: A unit of force equal to the force which will impart an acceleration of 1 foot per second per second to a mass of 1 ounce; equal to 0.0086409346485 newton.

Output shaft: The shaft that transfers motion from the prime mover to the driven machines.

Overarm: One of the flexible supports for the end of a milling-cutter arbor farthest from the machine spindle.

Overdrive: An automobile engine apparatus that lowers the gear ratio, thereby reducing fuel consumption.

Overfire draft: The air pressure in a boiler furnace during occurrence of the main flame.

Over gear: A gear train in which the angular velocity ratio of the driven shaft to driving shaft is greater than unity.

Overhead camshaft: A camshaft mounted above the cylinder head.

Overhead shovel: A tractor which digs with a shovel at

its front end, swings the shovel rearward overhead, and dumps the shovel at its rear end.

Overhead traveling crane: A hoisting machine with a bridge like structure moved on wheels along overhead trackage which is usually fixed to the building structure.

Overhead-valve engine: A four stroke-cycle internal combustion engine having its valves located in the cylinder head, operated by push rods that actuate rocker arms.

Overrunning clutch: An overrunning clutch transmits torque in one direction only and permits the driven shaft of a machine to free-wheel, or keep on rotating when the driver is stopped.

Overshot wheel: A horizontal-shaft water wheel with buckets around the circumference; the weight of water pouring into the buckets from the top rotates the wheel.

Over speed governor: A governor that stops the prime mover when speed is extreme.

Overspin: A rotating motion given to a ball when throwing or hitting it, used to give it extra speed or distance or to make it bounce awkwardly.

Oversquare engine: An engine with bore diameter greater than the stroke length.

Over steer: The propensity of an automotive vehicle to steer into a turn to a sharper degree than was intended by the driver; some-times causes the vehicle's rear end to swing out.

Overtone: One of the normal method of vibration of a vibrating system whose frequency is greater than that of the fundamental mode.

Oxygen mask: A mask that covers the nose and mouth and is used to manage oxygen.

Oxygen point: The temperature at which liquid oxygen and its vapor are in equilibrium, that is, the boiling point of oxygen, at standard atmospheric pressure.

P acking density: The number of storage cells per unit length, area, or volume of storage media.

Paddle wheel: A device used for the movement of shallow-draft vessels, consisting of a wheel with blades or floats in a circle, the wheel rotating in a plane parallel to the length of the vessel.

Pairing element: Either of two machine parts connected to permit motion.

Palpable coordinate: Generalized coordinate, which is explicitly included in the Lagrangian of the system.

Pancake engine: A compact engine with cylinders arranged radially.

Pan conveyor: A conveyor consisting of a series of pans.

Pan crusher: A device for grinding solid particles, in which one or more grinding wheels or grinders rotate in a pan containing the material to be ground.

Panel board: Drawing board with an adjustable outer frame that presses against the drawing paper to hold and stretch it.

Paper machine: A synchronized series of mechanical devices for converting a diluted suspension of cellulose fibers into a dry sheet of paper.

Parallel axis theorem: A theorem that states that the moment of inertia of a body about any given axis is the moment of inertia about a parallel axis passing through the center of mass, plus the moment of inertia that the body would have about this axis if the entire mass of the body were located at the center of mass.

Parallel linkage: Automotive steering system with a short idle arm parallel to the connecting rod.

Parking brake: In a motor vehicle, a brake that operates independently of the service brake and is applied after the vehicle has been stopped.

Parsons-stage steam turbine: A steam turbine having a reaction-type stage in which the pressure drop occurs partially across the stationary nozzles and partly across the rotating blades.

Particle dynamics: Study of the dependence of the motion of an individual material particle on external forces acting on it, in particular, electromagnetic and gravitational forces.

Particle energy: For a particle in potential - the sum of the kinetic energy of the particle and the potential energy.

Particle mechanics: The study of the motion of a single material particle.

Pascal: A unit of pressure equal to the pressure resulting from a force of 1 newton acting uniformly over an area of 1 square meter. Symbolized Pa.

Pass: The number of times that combustion gases are exposed to heat transfer surfaces in boilers (that is, single-pass, double-pass, and so on).

Passive solar system: A solar heating or cooling system that uses gravity, heat flux, or evaporation rather than mechanical devices to collect and transfer energy.

Pattern: A form designed and used as a model for making things.

Paver: Any of several machines which, moving along the road, carry and lay paving material.

Pawl: The driving link or holding link of a ratchet mechanism, permits motion in one direction only.

Peaucellier linkage: A mechanical linkage to convert circular motion exactly into straight-line motion.

Pebble mill: A device for grinding solid particles with a cylindrical or conical shell rotating on a horizontal axis and with grinding media such as flint, steel or porcelain balls.

Peck: A unit of volume used in the United States for measurement of solid substances, equal to 8 dry quarts, or 1/4 bushel, or 537.605 cubic inches, or 0.00880976754172 cubic meter.

Pedestal design: A robot design centered on the vertical axis of a central pedestal, in which the motion of any work piece is confined to a spherical working envelope.

Peep door: A small door in the firebox with a glass hole through which you can observe the combustion.

Pellet mill: A device for injecting feed in the form of particles, granules or paste into the holes of the roller, followed by compaction of the feed into a solid rod, which must be cut with a knife along the periphery of the roller.

Pelton wheel: Pulse hydraulic turbine, in which the pressure of the supplied water is converted into speed by means of several stationary nozzles, and then the water jets collide with the blades mounted on the rim of the wheel; generally limited to installations with high heads exceeding 500 feet (150 meters).

Pendulous gyroscope: A gyroscope whose axis of rotation is limited by a suitable weight to remain horizontal; it is the basis of one type of gyrocompass.

Pendulum saw: A circular saw that swings in a vertical arc for crosscuts.

Penetration ballistics: A section of terminal ballistics describing the movement and behavior of the missile during and after penetration into the target.

Penetration rate: The actual rate of penetration of drilling tools.

Penetration speed: The speed at which a drill can cut through rock or other material.

Pennyweight: A unit of mass equal to 1/20 troy ounce or to 1.55517384 grams; the term is employed in the United States and in England for the valuation of silver, gold, and jewels.

Perch: Also known as pole; rod. 1. A unit of length, equal to 5.5 yards, or 16.5 feet,

Percussion bit: A rock drilling tool with chisel like cutting edges that, when hitting the rock surface, drills a hole by spalling.

Percussion drill: A drilling rig that typically uses compressed air to drive a piston that strikes a series of blows against the shank of the drill rod or steel and the attached bit.

Percussion drilling: A drilling method in which hammer blows are transmitted by the drill rods to the drill bit.

Periodic motion: Any motion that repeats itself identically at regular intervals.

Peristaltic pump: A device for moving liquids due to the action of several evenly spaced rollers that rotate and compress a flexible tube.

Permanent axis: The axis of the greatest moment of inertia of a rigid body around which it can rotate in equilibrium.

Permanent benchmark: An easily identifiable, relatively constant, recoverable reference that is designed to maintain a constant altitude over an extended period of time with respect to the accepted coordinate system and is located where disturbances are considered insignificant.

Permanent gas: A gas at a pressure and temperature far from its liquid state.

Permanent set: Permanent plastic deformation of a structure or specimen after removal of the applied load.

Perpendicular axis theorem: A theorem that states that the sum of the moments of inertia of a flat plate relative to any two perpendicular axes in the plane of the plate is equal to the moment of inertia about an axis passing through their intersection perpendicular to the plate.

Phase: The type of state of a system, such as solid, liquid, or gas.

Phase diagram: A graph showing the pressures at which phase transitions between different states of a pure compound occur, as a function of temperature.

Philips hot air engine: A compact hot air engine that is a Philips Research Lab (Holland) design; it uses only one cylinder and piston, and operates at 3000 revolutions per minute,

Phonon friction: The friction that occurs when atoms close to a surface are set in motion by the sliding action of atoms on the opposite surface, and the mechanical energy required to slide one surface over another is thereby converted into vibrational energy of the atomic lattice (phonons). and turns into heat over time.

Physical testing: Determination of physical properties of materials based on observation and measurement.

Pico second: A unit of time equal to 10^{-12} second, or one-millionth of a microsecond (abbreviated ps).

Picowatt : A unit of power equal to 10^{-12} watt, or one-millionth of a microwatt. Abbreviated pW.

Pièze: A unit of pressure equal to 1 sthène per square meter, or to 1000 pascals.

Piezo meter: An apparatus for measuring fluid pressure in a system.

Pile driver: Hoist and movable steel frame designed for lifting piles and driving them into the ground.

Pile extractor: Pile hammer that hits the pile upward to loosen the grip and lift it off the ground. A vibratory hammer that loosens the pile with high-frequency shock.

Pile hammer: The heavy weight of a pile driver whose impact force depends on gravity and is used to drive piles into the ground.

Pillar crane: A crane whose mechanism can be rotated about a fixed pillar.

Pillar press: Punching press on two vertical racks; the drive shaft passes through the columns and the slider acts between them.

Pilot drill: A small drill to start a hole to ensure that a larger drill will run true to center.

Pinion: The smaller of a pair of gear wheels or the smallest wheel of a gear train.

Pint : Abbreviated pt. A unit of volume, used in the United States for measurement of liquid substances

Pin type mill: A solids grinder in which protruding pins on a rapidly rotating disc provide energy to break.

Piston: A sliding metal cylinder that reciprocates in a tubular housing, either moving against or moved by fluid pressure.

Piston blower: A piston-operated, positive-displacement air compressor used for stationary, automobile, and marine duty.

Piston corer: A steel pipe that is driven into the sediment by free fall and lead attached to the top end, and is capable of collecting undistorted vertical sections of sediment.

Piston displacement: The volume that a piston moves in a cylinder in one stroke is equal to the distance the piston travels multiplied by the inner cross section of the cylinder.

Piston drill: Heavy hammer drill, mounted either on a horizontal bar or on a short horizontal arm attached to a vertical column; Drills holes up to 6 inches (15 centimeters) in diameter. Also known as a piston drill.

Piston engine: A type of engine characterized by reciprocating motion of pistons in a cylinder.

Piston head: That part of a piston above the top ring

Piston pin: A cylindrical pin that connects the connecting rod to the piston. Also known as wrist pin

Piston pump: A pump in which motion and pressure is

applied to the fluid by a reciprocating piston in a cylinder. Also known as a piston pump.

Piston rod: The rod which is connected to the piston, and moves or is moved by the piston.

Piston skirt: That part of a piston below the piston pin bore.

Piston speed: The total distance a piston travels in a given time; usually expressed in feet per minute.

Piston valve: A cylindrical type of steam engine slide valve for admission and exhaust of steam.

Pitch acceleration: The angular acceleration of an aircraft or missile about its lateral, or Y, axis.

Pitch attitude: The position of an airplane, rocket, or other aircraft, related to the relationship between the longitudinal axis of the body and the selected baseline or plane, as viewed from the side.

Pitch axis: A lateral axis through an aircraft, missile, or similar body, about which the body pitches.

Pitching moment: A moment about a lateral axis of an aircraft, rocket, or air-

foil.

Pivot: A short, pointed shaft forming the center and fulcrum on which something turns, balances, or oscillates.

Pivot-bucket conveyor-elevator: Bucket conveyor with overlapping swivel buckets on long pitch roller chains; buckets are always leveled, except when stumbling to unload material.

Plain turning: Lathe operations involved when machining a workpiece between centers.

Planar linkage: A linkage that involves motion in only two dimensions.

Planck function: The negative of the Gibbs free energy divided by the absolute temperature.

Plane lamina: A body whose mass is concentrated in a single plane.

Plane of departure: Vertical plane containing the path of a projectile as it leaves the muzzle of the gun.

Plane of fire: Vertical plane containing the gun and the target, or containing a line of site.

Plane of maximum shear stress: Any of two planes that lie on opposite sides and at angles of 45 to the maximum principal stress axis and that are parallel to the intermediate principal stress axis.

Plane of yaw: The plane determined by the tangent to the trajectory of a projectile in flight and the axis of the projectile.

Plan equation: The mathematical statement that horsepower plan/33,000, where p mean effective pressure (pounds per square inch), l length of piston stroke (feet), a net area of piston (square inches), and n number of cycles completed per minute.

Planer: A machine for forming long, flat or flat surfaces with contours by reciprocating a workpiece under a stationary single-point tool or tools.

Plane strain: Deformation of a body, in which the displacements of all points of the body are parallel to a given plane, and the values of these displacements do not depend on the distance perpendicular to the plane.

Plane stress: A state of stress in which two of the principal stresses are always parallel to a given plane and are constant in the normal direction.

Planetary gear train: A meshing gear assembly consisting of a center gear, a co-axial inner or ring gear, and one or more idler gears supported by a rotating carrier.

Planet carrier: A fixed member in a planetary gear train that contains the shaft upon which the planet pinion rotates.

Planet gear: A pinion in a planetary gear train.

Planet pinion: One of the gears in a planetary gear train that meshes with and revolves around the sun gear.

Planishing: Smoothing the surface of a metal by a rapid series of overlapping, light hammerlike blows or by rolling in a planishing mill.

Plasma arc cutting: Metal cutting by melting a localized area with an arc followed by removal of metal by high velocity, high temperature ionized gas.

Plastic: Displaying, or associated with, plasticity.

Plastic collision: A collision in which one or both colliding bodies are subjected to plastic deformation and mechanical energy is dissipated.

Plastic deformation: Permanent change in the shape or size of a solid without fracture as a result of the application of prolonged stress exceeding the elastic limit.

Plasticity: The property of a solid, in which it undergoes an irreversible change in shape or size under the influence of a stress exceeding a certain value, is called the yield strength.

Plastic viscosity: Plasticity, at which the rate of deformation of a body subjected to stresses exceeding the yield point is a linear function of the stress.

Plate cam: A flat, open cam that imparts a sliding motion.

Plate coil: The heat transfer device is made of two metal sheets fastened together, one or both of the plates are embossed, forming passages between them for the flow of a heating medium or cooling medium.

Plate conveyor: Conveyor with a series of steel plates as the carrier medium; each plate is a short groove, all with slight overlap to form a hinge band and attached to one central chain or two side chains; Chains connect to rollers running on a steel angle and transmit power from drive heads installed at intermediate points, and sometimes also at the head or tail ends.

Plate fin exchanger: A heat transfer device consisting of a package or layers, each of which consists of a corrugated rib between flat metal sheets, insulated on both sides by channels or rods that form passages for the flow of liquids.

Plate modulus: The ratio of the stress component T_{xx} in an isotropic, elastic body obeying a generalized Hooke's law to the corresponding strain component S_{xx}, when the strain components S_{yy} and S_{zz} are 0; the sum of the Poisson ratio and twice the rigidity modulus.

Plate type exchanger: Heat-exchange device similar to a plate-and-frame filter press; fluids flow between the frame-held plates, transferring heat between them.

Platform conveyor: One or two-lane conveyor with steel or hardwood plates forming a continuous platform on which loads are placed.

Play: Free or unobstructed movement of an object, such as movement between poorly fitted or worn parts.

Plenum blower assembly: In an automotive air conditioning system, the assembly through which air passes to the evaporator or heater core.

Plenum system: A heating or air conditioning system in which air is forced through a plenum chamber for distribution to ducts.

Pli: A unit of line density (mass per unit length) equal to 1 pound per inch, or approximately 17.8580 kilograms per meter.

Plug valve: A valve with a plug that has an opening through which fluid flows and that can be rotated 90 ° to operate in an open or closed position. Also known as plug valve.

Plunge: For setting the horizontal crosshair of the odolite in the direction of the slope when setting the slope between two points of a known level.

Plunge grinding: Grinding in which the wheel moves radially towards the working plunger pump: a piston pump in which the packing is on a stationary housing rather than on a moving piston.

Pneumatic atomizer: An atomizer that uses compressed air to produce drops in the diameter range of 5-100 micrometers.

Pneumatic controller: A device for mechanically moving another device (for example, a valve stem), the action of which is controlled by changes in the pneumatic pressure connected to the controller.

Pneumatic control valve: A valve in which the force of compressed air on a diaphragm opposes the force of a spring to control the area of the fluid flow hole.

Pneumatic conveyor: A conveyor that transports dry, free-flowing, granular material in the form of a slurry or cylindrical carrier through a pipe or duct using a high-speed air flow or vacuum pressure generated by an air compressor.

Pneumatic drill: Compressed-air drill worked by reciprocating piston, hammer action, or turbo drive.

Pneumatic drilling: Drilling a well using air or gas instead of conventional drilling fluid as the circulating medium; rotary drilling device.

Pneumatic hammer: A hammer in which compressed air is used to create an impact blow. Also known as an air hammer; jackhammer.

Pneumatic riveter: Riveting machine with a fast reciprocating piston driven by compressed air.

Poinsot motion: The motion of a rigid body with a fixed point in space and with zero torque or moment acting on the body relative to the fixed point.

Poinsot's central axis: A line passing through a rigid body, which is parallel to the vector sum F of the system of forces acting on the body, and which is located so that the system of forces is equivalent to a force F applied anywhere along the line, plus a pair whose torque is equal to the component of the total torque T applied by the system in direction F.

Poinsot's method: A method of describing Poinsot's motion using a geometric structure in which an ellipsoid of inertia rolls on a constant plane without slipping.

Point-blank range: Distance to a target that is so short that the trajectory of a bullet or projectile is practically a straight, rather than a curved, line.

Point of contraflexure: A point at which the direction of bending changes. Also known as point of inflection.

Point of fall: The point in the curved path of a falling projectile that is level with the muzzle of the gun.

Poisson ratio: Ratio of transverse compressive strain to elongation strain when a bar is stretched by forces applied to its ends and parallel to the bar axis.

Polar timing diagram: A diagram of the events of an engine cycle relative to crank shaft position.

Pole lathe: A simple lathe in which the work is rotated by a cord attached to a treadle.

Polhode: For a rotating rigid body not subject to an external torque, a closed curve drawn on an ellipsoid of inertia by intersecting with this ellipsoid an axis parallel to the angular velocity vector and passing through the center.

Polishing roll: Roll or series of rolls on a plastic mold; has polished chrome surfaces; It is used to obtain a smooth surface on a plastic sheet when extruded.

Polytropic process: An expansion or compression of a gas in which the quantity PV^n is held constant, where P and

V are the pressure and volume of the gas, and n is some constant(Polytropic Index)

Pop action: The action of a safety valve as it opens under steam pressure when the valve disk is lifted off its seat.

Poppet valve: A cam-operated or spring-loaded reciprocating engine mushroom-type valve used for control of admission and exhaust of working fluid.

Popping pressure: In compressible fluid service, the inlet pressure at which a safety valve disk opens.

Porcupine boiler: A boiler with dead end tubes projecting from a vertical shell.

Portal crane: A jib crane carried on a four legged portal built to run on rails.

Positioning: Tool function associated with the manipulation of the workpiece in relation to working tools.

Positioning time: The time required to move a machining tool from one coordinate position to the next.

Positive acceleration: Accelerating force in an upward direction or direction, such as from bottom to top or from seat to head. Acceleration in the direction of the application of this force.

Positive clutch: A clutch designed to transmit torque without slip.

Positive-displacement compressor: A compressor that holds successive volumes of fluid in an enclosed space in which the pressure of the fluid increases as the volume of the enclosed space decreases.

Positive-displacement pump: A pump in which a measured quantity of liquid is entrapped in a space, its pressure is raised, and then it is delivered; for example, a reciprocating piston-cylinder or rotary-vane, gear, or lobe mechanism.

Positive draft: The pressure in the furnace or gas paths of the steam generator is higher than atmospheric.

Positive motion: Motion transferred from one machine part to another without slippage.

Positive temperature coefficient: A condition in which the resistance, length, or some other characteristic of a substance increases with increasing temperature.

Post brake: The brake, sometimes mounted on a steam coiler or traction vehicle, consists of two vertical struts mounted on either side of the drum that operate on braking tracks bolted to the drum cheeks.

Pot die forming: Forming sheet or plate metal through a hollow die by the application of pressure which causes the work piece to assume the contour of the die.

Potential energy: The capacity to do work that a body or system has by virtue of its position or configuration.

Potential temperature: The temperature that a compressible fluid could reach if it were adiabatically compressed or expanded to a standard pressure, typically 1 bar.

Pound: A unit of mass in the English absolute system of units, equal to 0.45359237 kilogram.

Pound per square foot: A unit of pressure equal to the pressure resulting from a 1 pound force uniformly applied over an area of 1 square foot.

Pound per square inch: A unit of pressure equal to the pressure resulting from a 1 pound force uniformly applied over an area of 1 square inch.

Pounds per square inch absolute: Absolute thermodynamic pressure, measured in pound-force applied to an area of 1 square inch.

Pounds per square inch gage: The gage pressure, measured by the number of pounds force exerted on an area of 1 square inch.

Powder clutch: A type of electromagnetic disc clutch in which the space between the clutch elements is filled with dry fine magnetic particles; the application of a magnetic field unites the particles, creating frictional forces between the elements of the clutch.

Power-actuated pressure relief valve: A pressure relief valve connected to and controlled by a device which utilizes a separate energy source.

Power brake: An automobile brake with a vacuum in the engine intake manifold, used to increase atmospheric pressure on a piston that is actuated by the brake pedal.

Power car: A railroad car with equipment for furnishing heat and electric power to a train.

Power control valve: A safety relief device operated

by a power-driven mechanism rather than by pressure.

Power drill: A motor-driven drilling machine.

Power-driven: A component or piece of equipment that is moved, rotated, or driven by electrical or mechanical energy, as in power driven fans or power driven turrets.

Power package: A complete engine and its accessories, designed as a single unit for quick installation or removal.

Power plant: Any device that converts any form of energy into electrical energy, such as a hydroelectric or steam generator, a diesel-electric locomotive engine, or a nuclear power plant.

Power saw: A power operated woodworking saw, such as a bench or circular saw.

Power shovel: A power operated shovel that carries a short boom on which rides a movable dipper stick carrying an open-topped bucket; used to excavate and remove debris.

Power steering: A steering system for a moving vehicle in which an auxiliary power source assists the driver by providing the main force required to steer the support wheels.

Power stroke: The stroke in an engine during which pressure is applied to the piston by expanding steam or gases.

Power train: The part of the vehicle connecting the engine to the propeller or drive axle; may include drive shaft, clutch, transmission and differential.

Poynting effect: The effect of torsion of a very long cylindrical rod on its length.

Poynting's law: A special case of the Clapeyron equation, in which the liquid is removed as quickly as it is formed, so that its volume can be ignored.

Prandtl number: Dimensionless number used in the study of forced and free convection, equal to the dynamic viscosity multiplied by the specific heat at constant pressure divided by the thermal conductivity.

Pre breaker: Device used to break down large masses of solids prior to feeding them to a crushing or grinding device.

Precession: The angular velocity of the axis of rotation of a rotating rigid body resulting from the action of external moments on the body.

Precessional torque: A torque which causes a rotating body to precess.

Precision grinding: Machine grinding to specified dimensions and low tolerances.

Pre combustion chamber: A small chamber in front of the main combustion space of a turbine or piston engine in which combustion is initiated.

Pre cooler: A device for reducing the temperature of a working fluid before it is used by a machine.

Pre heater: A device for preliminary heating of a material, substance, or fluid that will undergo further use or treatment by heating.

Pre ignition: Ignition of the charge in the cylinder of an internal combustion engine before ignition by the spark.

Preset tool: A machine tool that is used to set an initial value of a parameter controlling another device.

Press: Any of a variety of machines that apply pressure to a workpiece, cut or shape material under pressure, compress

the material, or release fluid.

Press slide: The reciprocating member of a power press on which the punch and upper die are fastened.

Pressure: A type of stress which is exerted uniformly in all directions; its measure is the force exerted per unit area.

Pressure angle: The angle that a line of force forms with a line at right angles to the centerline of the two gears at the pitch points.

Pressure bar: A rod that holds the edge of a metal sheet during pressing operations such as punching, punching or forming, and prevents the sheet from warping or warping.

Pressure coefficient: The ratio of the partial pressure change to the temperature change under given conditions, usually at constant volume.

Pressure containing member: The part of the pressure relief device that is in direct contact with the pressurized medium in the protected vessel.

Pressure plate: The part of an automobile disk clutch that presses against the flywheel.

Pressure regulator: Open-close device used on the vent of a closed, gas-pressured system to maintain the system pressure within a specified range.

Pressure relief device: In pressure vessels, a device designed to open in a controlled manner so that the internal pressure of a component or system does not exceed a predetermined value, i.e. a safety valve.

Pressure-relief valve: A valve that relieves pressure beyond a specified limit and resets when returned to normal operating conditions.

Pressure retaining member: That part of a pressure relieving device loaded by the restrained pressurized fluid.

Pressure system: Any system of pipes, vessels, tanks, reactors, and other equipment,

Pressure tank: A pressurized tank into which timber is inserted for impregnation with preservative.

Pressure transducer: A component of an instrument that senses the pressure of a fluid and provides an electrical signal related to the pressure.

Pressure travel curve: Curve showing pressure plotted against the travel of the projectile within the bore of the weapon.

Pressurize: To keep normal atmospheric pressure in a chamber subjected to high or low external pressures.

Prevost's theory: The theory that a body constantly exchanges heat with its surroundings emits an amount of energy that is independent of its surroundings, and increases or decreases its temperature depending on whether it absorbs more radiation than it emits, or vice versa.

Primary air: That portion of the combustion air introduced with the fuel in a burner.

Primary breaker: A machine which takes over the work of size reduction from blasting operations, crushing rock to maximum size of about 2-inch (5-centimeter) diameter; may be a gyratory crusher or jaw breaker. Also known as primary crusher.

Primary creep: The initial high strain rate region in a material subjected to sustained stress.

Primary phase: The only crystalline phase that can exist in equilibrium with a given liquid.

Primary phase region: On a phase diagram, the locus of all compositions having a common primary phase.

Primary stress : The component of normal or shear stress in a solid material that results from an applied load is in equilibrium and is not self-limiting.

Prime mover: A power plant component that converts energy from a thermal or pressure form to a mechanical form.

Priming: In the boiler - excessive entrainment of fine water particles along with the steam due to insufficient steam space, improper boiler design or incorrect operating conditions.

Priming pump: A device on motor vehicles and tanks, providing a means of injecting a spray of fuel into the engine to facilitate starting.

Principal axis of strain: One of the three axes of a body that were mutually perpendicular before deformation.

Principal axis of stress: One of the three mutually perpendicular axes of a body that are perpendicular to the principal planes of stress.

Principal strain: The elongation or compression of one of the principal axes of strain relative to its original length.

Principal stress: A stress occurring at right angles to a principal plane of stress.

Principle of dynamical similarity: The principle according to which two physical systems that are geometrically and kinematically similar at a given moment and are physically similar in structure will retain this similarity at later corresponding points in time if and only if the Froude number 1 for each independent type of force has identical values of two systems.

Principle of least action: The principle according to which for a system, the total mechanical energy of which is conserved, the trajectory of the system in the configuration space is a path that makes the action value stationary relative to nearby paths between the same configurations and for which the energy has the same constant. value.

Principle of virtual work: The principle that the total work done by all forces acting on a system in static equilibrium is zero for any infinitesimal displacement from equilibrium

Prism joint: A robotic articulation that has only one degree of freedom, in sliding motion only.

Profiling machine :A machine used for milling irregular profiles; the cutting tool is guided by the contour of a model.

Prony brake: An absorbent dynamometer that applies a frictional load to the output shaft by means of wooden blocks, flexible tape, or other frictional surface.

Proof load: A predetermined test load, greater than the service load, to which a specimen is subjected before acceptance for use.

Proof resilience: The tensile strength required to stretch an elastomer from zero elongation to breaking point, expressed in pounds-feet per cubic inch of original size.

Proof stress: The stress that causes a specified amount of permanent deformation in a material.

Propeller: A bladed device that rotates on a shaft to produce a useful thrust in the direction of the shaft axis.

Propeller efficiency: The ratio of the thrust transmitted by the propeller to the shaft power transmitted by the motor to the propeller.

Propeller fan: An axial-flow blower, with or without a casing, using a propeller-type rotor to accelerate the fluid.

Propeller slip angle: The angle between the plane of the blade face and its direction of motion.

Propeller tip speed: The speed in feet per minute swept by the propeller tips.

Propeller turbine: A form of reactive-type hydraulic turbine using an axial-flow propeller rotor.

Propeller wind mill: A windmill that extracts wind energy from horizontal air movements to rotate the propeller blades.

Propulsion : The process of causing a body to move by exerting a force against it.

Propulsion system: For a vehicle moving in a fluid medium, such as an airplane or ship, a system that produces the required change in the amount of motion in the vehicle by changing the speed of air or water passing through a propulsion device or engine; in the case of a rocket vehicle operating without a fluid, the required change in momentum is achieved by using some of the propellant's own mass, called a propellant.

Psychromatic ratio: Ratio of the heat transfer coefficient to the product of the mass-transfer coefficient and humid heat for a gas-vapor system; used in calculation of humidity or saturation relationships.

Psychrometric chart: A graph each point of which represents a specific condition of a gas-vapor system (such as air and water vapor) with regard to temperature (horizontal scale) and absolute humidity (vertical scale); other characteristics of the system, such as relative humidity, wet-bulb temperature, and latent heat of vaporization, are indicated by lines on the chart.

Psychrometric formula: Semi-empirical ratio giving the vapor pressure according to the readings of the barometer and psychrometer.

Psychrometric tables: The tables are based on a psychrometric formula and are used to derive vapor pressure, relative humidity

and dew point from wet bulb and dry bulb temperatures.

Puff: A small explosion within a furnace due to combustion conditions.

Pug mill: A machine for mixing and tempering a plastic material by the action of blades revolving in a drum or trough.

Puller: Link-driven chain or cable winch for lifting or pulling at any angle, which has a reversible ratchet in the lever to allow short travel for both pulling and loosening, and which holds loads with a Weston-type friction brake or a removable ratchet.

Pulley top: A top with a long shank used to tap setscrew holes in pulley hubs.

Pull in torque: The largest steady torque with which a motor will attain normal speed after accelerating from a standstill.

Pull-out torque: The largest torque under which a motor can operate without sharply losing speed.

Pull strength: A unit in tensile testing; the bond strength in pounds per square inch.

Pulsometer: A simple,

lightweight pump in which steam forces water out of one of two chambers alternately.

Pulverizer: Device for breaking down of solid lumps into a fine material by cleavage along crystal faces.

Pump bob: A device such as a crank that converts rotary motion into reciprocating motion.

Pumping loss: Power consumed in purging a cylinder of exhaust gas and sucking in fresh air instead.

Punch press: A press consisting of a frame in which slides or rams move up and down, of a bed to which the die shoe or bolster plate is attached, and of a source of power to move the slide.

Purge meter interlock: A meter to maintain airflow through a boiler furnace at a specific level for a definite time interval; ensures that the proper air-fuel ratio is achieved prior to ignition.

Push bar conveyor: A type of chain conveyor in which two endless chains are cross-linked at regular intervals by pushers that move the load along a fixed base or chute of the conveyor.

Push bench: A machine used for drawing tubes of moderately heavy gage by cupping metal sheet and applying pressure to the inside bottom of the cup to force it through a die.

Push nipple: A short length of pipe used to connect sections of cast iron boilers.

Push rod: A stem, like in an internal combustion engine, which is driven by a cam to open and close the valves.

Pyrometry: The science and technology of measuring high temperatures.

Q

Quadrant angle of fall: The vertical acute angle at the level point, among the horizontal and the line of fall of a projectile.

Quadricycle: A four wheeled human powered land vehicle, typically propelled by the action of the rider's feet on the pedals.

Quantity meter: A type of fluid meter used to calculate volume of flow.

Quarrying machine: Any machine used to drill holes or cut tunnels in native rock, such as a gang drill or tunneling machine; most commonly, a small locomotive bearing rock-drilling equipment operating on a track.

Quart: Abbreviated qt. A unit of volume used for measurement of liquid substances in the United States,

Quarter: A unit of mass is use in the United States, equal to 1/4 short ton.

Quartering machine: A machine that bores parallel holes simultaneously in such a way that the center lines of adjacent holes are 90 apart.

Quarter-turn drive: A belt drive connecting pulleys whose axes are at right angles

Quick return: A tool used in a reciprocating machine to make the return stroke faster than the power stroke.

Quill drive: A drive in which the motor is mounted on a non-rotating hollow shaft surrounding the driving-wheel axle; pins on the armature mesh with spokes on the driving wheels.

Quill gear: A gear climbed on a hollow shaft.

Rack and pinion: A gear train consisting of a toothed rack that meshes with a pinion.

Rack-and-pinion steering: A steering system in which rotation of the pinion at the end of the steering column moves a pinion (rack) left or right to transmit steering movements.

Radial band pressure: The pressure exerted on the rotating band of the gun barrel wall and, therefore, on the projectile wall in the band slot as a result of the rifle's thread engraving the band.

Radial bearing: A rolling contact bearing in which the direction of action of the transmitted load is radial to the shaft axis.

Radial draw forming: A metal working method in which tangential tension and radial compression are applied gradually and simultaneously.

Radial drill: A drilling machine in which the drilling spindle can be moved along a horizontal arm, which itself can rotate around a vertical column.

Radial drilling: The drilling of several holes in one plane, all radiating from a common point.

Radial engine: An engine characterized by radially arranged cylinders at equiangular intervals around the crankshaft

Radial-flow: Having the fluid working substance flowing along the radii of a rotating tank

Radial flow turbine: A turbine in which the gases flow primarily in a radial direction.

Radial force: In machining, the force acting on the cutting tool in the opposite direction to the depth of cut.

Radial heat flow: The heat flow between two coaxial cylinders is maintained at different temperatures; used to measure the thermal conductivity of gases.

Radial load: The load perpendicular to the bearing axis.

Radial locating: One of three problems with tool placement to maintain the desired ratio between work piece, cutter and machine body; the other two location tasks are concentric and flat localization.

Radial motion: Movement

in which a body moves along a line connecting it to an observer or a reference point; for example, the movement of stars that move towards or away from Earth without changing their apparent position.

Radial rake: The angle between the cutter tooth surface and the radial line passing through the cutting edge in a plane perpendicular to the cutter axis.

Radial saw: An electric saw with a circular blade suspended from a traverse mounted on a pivot arm.

Radial stress: Tangential stress at the periphery of an opening.

Radial velocity: The component of the velocity of a body that is parallel to a line from an observer or reference point to the body.

Radial wave equation: Solutions to spherical symmetry wave equations can be found by separating variables; the ordinary differential equation for the radial part of the wave function is called the radial wave equation.

Radiant super heater: Superheater designed to transfer heat from combustion products to steam, mainly through radiation

Radiant type boiler: A water-tube boiler, in which the boiler tubes form the border of the furnace.

Radiation loss: Boiler heat loss to the atmosphere by conduction, radiation, and convection.

Radiator temperature drop: In internal combustion engines, the temperature difference between the coolant entering and leaving the radiator.

Radioactive heat: Heat produced within a medium as a result of absorption of radiation from decay of radioisotopes in the medium, such as thorium-232, potassium-40, uranium-238, and uranium-235.

Radius cutter: A formed milling cutter with teeth ground to produce a radius on the work piece.

Radius of gyration: The square root of the ratio of the moment of inertia of a body about a given axis to its mass.

Railroad jack: A hoist used for lifting locomotives.

Ram: A plunger, weight, or other guided structure for exerting pressure or drawing something by impact.

Ram effect: The increased air pressure in a jet engine or in the manifold of a piston engine, due to ram.

Ramsay Young method: A method of measuring the vapor pressure of a liquid, in which the bulb of a thermometer is surrounded by cotton soaked in liquid, and the pressure measured by a pressure gauge is reduced until the thermometer readings become stable.

Ramsay-Young rule: An empirical relationship that states that the ratio of the absolute temperatures at which two chemically similar liquids have the same vapor pressure is independent of that vapor pressure.

Ram type turret lathe: A horizontal turret lathe in which the turret is mounted on a ram or slide which rides on a saddle.

Random vibration: A varying force acting on a mechanical system which may be considered to be the sum of a large number of irregularly timed small shocks; induced typically by aerodynamic turbulence, airborne noise from rocket jets, and transportation over road surfaces.

Range deviation: Distance by which a projectile strikes beyond, or short of, the target; the distance as measured along the gun-target line or along a line parallel to the gun-target line.

Rankine cycle: Ideal thermodynamic cycle, consisting of heat input at constant pressure, isentropic expansion, heat removal at constant pressure and isentropic compression; It is used as an ideal standard for the operation of heat engines and heat pumps operating with condensed steam as a working medium, for example steam power plants. Also known as steam cycle.

Rankine efficiency: The efficiency of an ideal Rankine cycle engine under specified steam temperature and pressure conditions.

Rankine Hugoniot equations: Equations derived from the laws of conservation of mass, momentum and energy that relate the shock wave velocity and pressure, density and enthalpy of the transmitting fluid before and after the shock wave has passed.

Rankine temperature scale: A scale of absolute temperature; the temperature in degrees Rankine (R) is equal to 9/5 of the temperature in kelvins and to the temperature in degrees Fahrenheit plus 459.67.

Rapid traverse: A machine tool mechanism which rapidly repositions the work piece while no cutting takes place.

Ratchet coupling: A coupling between two shafts that uses a ratchet that allows the driven shaft to rotate in only one direction and also allows the driven shaft to overtake the drive shaft.

Rated capacity: The maximum capacity a boiler is rated for is measured in pounds of steam per hour delivered under specified pressure and temperature conditions.

Rated engine speed: Engine speed is listed as maximum allowed for continuous reliable operation.

Rated horsepower: Normal maximum, allowable, continuous power output of an engine, turbine engine, or other prime mover.

Rated load: The maximum load a machine is designed to carry.

Rate gyroscope: A gyroscope suspended on only one suspension, the bearings of which form its output axis and which is held by a spring; rotation of the gyroscope frame around an axis perpendicular to both the rotation axis and the output axis causes the suspension to precession inside the bearings, proportional to the rotation speed.

Rate integrating gyroscope: A single degree of freedom gyro having primarily viscous restraint of its spin axis about the output axis;

Rate of change of acceleration: Time rate of change of acceleration; this rate is a factor in the design of some items of ammunition that undergo large accelerations.

Ratio of expansion: The ratio of the volume of steam in the engine cylinder when the piston is at the end of its stroke to when the piston is in the cut-off position.

Rayleigh line: A straight line connecting the points corresponding to the initial and final states on the plot of pressure versus specific volume for a substance subjected to a shock wave.

Rayleigh number: Dimensionless number used in the study of free convection, equal to the product of the Grashof number and the Prandtl number.

Rayleigh's dissipation function: A function that enters the equations of motion of a system experiencing small fluctuations and represents the frictional forces proportional to the velocities; is given by a positive definite quadratic form of the derivatives of coordinates with respect to time.

Rayleigh wave: A wave propagating over the surface of a solid; particle trajectories are ellipses in planes normal to the surface and parallel to the direction of propagation.

Reach rod: Rod movement in a link used to transfer motion from the reversing rod to the lift shaft.

Reaction turbine: A prime mover for power generation using the principle of steady flow of acceleration of a fluid in which nozzles are mounted on a moving element.

Reaction wheel: A device capable of storing angular momentum that can be used in a spacecraft to provide torque to achieve or maintain a given orientation.

Real gas: A gas, as considered from the viewpoint in which deviations from the ideal gas law, resulting from interactions of gas molecules, are taken into account. Also known as imperfect gas.

Receiving station: A location or device on conveyor systems where bulk material is loaded or otherwise taken onto a conveyor.

Recessed tube wall: The wall of the boiler furnace, which has holes for partial ingress of radiant flue gases into the pipes of the water walls.

Reciprocal strain ellipsoid: In the theory of elasticity, an ellipsoid of a certain shape and orientation, which, under uniform deformation, turns into a set of orthogonal diameters of a sphere.

Reciprocating compressor: A positive displacement compressor having one or more cylinders, each of which is equipped with a piston driven by a crankshaft through a connecting rod.

Reciprocating-plate feeder: Reciprocating tray used to feed abrasive materials such as coal dust into processing plants.

Reciprocating screen: Horizontal solids-separation screen (sieve) oscillated back and forth by an eccentric gear; used for solids classification.

Recirculating ball steering: Steering system that transmits steering movements using steel balls placed between the worm gear and the nut.

Recovery: Returning the body to its original size after stress, possibly over a significant period of time.

Recovery vehicle: Special purpose vehicle equipped with a winch, hoist or boom for evacuating vehicles.

Rectilinear motion: Continuous change of body position so that each particle of the body moves in a straight line.

Redler conveyor: A conveyor in which material is pulled through a channel by frame or U-shaped impellers that move the material in which they are immersed because the resistance to slip through the element is greater than the resistance to the channel walls.

Reduced pressure: The ratio of the pressure of a substance to its critical pressure.

Reduced temperature: The ratio of the temperature of a substance to its critical temperature.

Reduced value: The actual value of the quantity divided by the value of that quantity at the tipping point. Also known as diminished property.

Reduced volume: The ratio of the specific volume of a substance to its critical volume.

Reduction gear: A gear train which lowers the output speed.

Redundancy: A statically indeterminate structure.

Refractory-lined firebox boiler: Horizontal fire tube boiler with the front part of the shell located above the refractory furnace; the rear part of the shell contains pipes of the first passage, and the pipes of the second passage are located in the upper part of the shell.

Refrigerated truck: Insulated truck equipped and used as a refrigerator for the transport of fresh perishable or frozen food.

Refrigeration: The cooling of a space or substance below the environmental temperature.

Refrigeration condenser: A vapor condenser in a refrigeration system where the refrigerant is liquefied and gives off heat to the environment.

Refrigeration cycle: A sequence of thermodynamic processes in which heat is removed from a cold body and removed to a hot one.

Refrigeration system: A closed flow system in which refrigerant is compressed, condensed and expanded to cool at a lower temperature level and remove heat at a higher level to remove heat from the controlled space.

Refrigerator: An insulated, cooled compartment.

Refrigerator car : An insulated freight car constructed and used as a refrigerator.

Regelation: Phenomenon in which ice (or any substance which expands upon freezing) melts under intense pressure and freezes again when this pressure is removed; accounts for phenomena such as the slippery nature of ice and the motion of glaciers.

Regeneration system: A system within a gas turbine that recovers waste heat from the turbine exhaust and uses it for the compression cycle.

Regenerative air heater: An air heater in which heat transfer elements are alternately exposed to warm gases and air.

Regenerative cycle: A motor cycle in which low grade heat, which would normally be lost, is used to improve cyclic efficiency.

Regenerative pump: A rotating vane device that uses a combination of mechanical impulse and centrifugal force to create a high fluid head at low volumes. Also known as a turbine pump.

Rehabilitation engineering: Using technology to maximize the independence of people with disabilities by providing assistive devices to compensate for the disability.

Reheating: A process in which a gas or steam is reheated after partial isentropic expansion to reduce moisture content. Also known as reheating.

Relative momentum: The momentum of a body in a frame of reference in which another given body is fixed.

Relative motion: Continuous change in body position relative to a second body or a fixed reference point. Also known as apparent motion.

Relative velocity: The velocity of a body with respect to a second body; that is, its velocity in a reference frame where the second body is fixed.

Relaxation: Relief of stresses in a deformed material due to creep. A decrease in elastic resistance in an elastic medium under the action of an applied stress, leading to permanent deformation.

Release : Mechanical arrangement of parts to hold or release a device or mechanism as needed.

Release adiabat: A curve or locus of points that defines a sequence of states through which a mass that has been shaken to a high pressure condition passes, returning monotonically to zero pressure.

Relief: A pass made by cutting off one side of the center of the tailstock so that the facing or parting tool can be advanced towards or near the center of the part. Providing clearance around the cutting edge by removing tool material.

Relief angle: The angle between a relieved surface and a tangential plane at a cutting edge.

Relief frame: Frame located between the steam engine spool and the steam box cover; reduces the pressure on the valve and thus reduces friction.

Relieving: Abrasion of an embossed metal surface to reveal the color of the base metal on elevations or light areas of the surface.

Remaining velocity: Speed of a projectile at any point along its path of fire.

Remote center compliance: A compatible device that allows a part gripped by a robot or other automatic equipment to rotate around the tip of a robot end actuator or move without rotating when pressed, thereby simplifying mechanical assembly of parts.

Repair: To restore an unusable condition by replacing parts, components or assemblies.

Repeated load: A force applied repeatedly, causing a change in magnitude, and sometimes in the sense of internal forces.

Replica master: A robot-like machine whose movements are duplicated by another robot when the machine is moved by a human operator.

Repulsion: A force that tends to increase the distance between two bodies that have the same electrical charges, or a force between atoms or molecules at very short distances that separates them.

Resealing pressure: The inlet pressure at which the leak stops when the pressure relief valve is closed.

Resilience : The ability of a deformed body, due to its high yield point and low modulus of elasticity, to recover its size and shape after deformation.

Resisting moment: The moment created by the internal tensile and compressive forces, which balances the external bending moment of the beam.

Resonance vibration: Forced vibration, in which the frequency of the disturbing force is very close to the natural frequency of the system, so that the vibration amplitude is very large.

Resultant rake: The angle between the surface of the cutting tooth and the axial plane through the point of the tooth, measured in a plane

perpendicular to the cutting edge.

Retarder: A braking device used to control the speed of railroad cars moving along the classification tracks in a hump yard.

Retarding conveyor: Any type of conveyor used to restrict the movement of bulk materials, packages or items where the slope is such that the material being conveyed tends to propel the conveying medium.

Return connecting rod : A connecting rod with a connecting rod end on the same side of the crosshead as the cylinder.

Return-flow burner: Mechanical oil diffuser in the boiler furnace, which regulates the amount of oil burned by recirculating part of the oil to the storage point.

Return idler: A roller, or roller under covers, along which the conveyor belt moves after the load it was carrying has been dropped.

Reverse Brayton cycle: A Brayton cycle that is driven in reverse direction is known as the reverse Brayton cycle. Its purpose is to move heat from colder to hotter body, rather than produce work. This cycle is also known as the gas refrigeration cycle or Bell Coleman cycle.

Reverse Carnot cycle: An ideal thermodynamic cycle consisting of the processes of the Carnot cycle reversed and in reverse order, namely, isentropic expansion, isothermal expansion, isentropic compression, and isothermal compression.

Reverse pitch: Propeller blade pitch that creates reverse thrust.

Reversible engine: An ideal engine which carries out a cycle of reversible processes.

Reversible path: The path followed by a thermodynamic system, so that its direction of motion can be changed at any point by an infinitely small change in external conditions; thus, the system can be considered to be in equilibrium at all points along the path.

Reversible-pitch propeller: Adjustable pitch propeller type; variable or constant speed, it has provisions for decreasing the step to zero and above, to a negative step range.

Reversible process: An ideal thermodynamic process that can be reversed exactly by introducing an infinitesimal change in external conditions. Also known as quasi-static process.

Reversible steering gear: A steering gear for a vehicle that allows shock loads and wheel deflection to pass through the system and be felt in the steering.

Revolute joint: A robotic articulation consisting of a pin with one degree of freedom.

Revolution: The motion of a body around a closed orbit.

Revolution per minute: A unit of angular velocity equal to the constant angular velocity of a body that rotates 360 (2 radians) so that each point of the body returns to its original position in 1 minute. Abbreviated rpm.

Revolution per second: A unit of angular velocity equal to the uniform angular velocity of a body that rotates 360 (2 radians) so that each point of the body returns to its original position in 1 second.

Revolving block engine: Any of a variety of engines combining reciprocating piston motion with rotary motion of the entire cylinder block.

Revolving shovel: A crawler-mounted or rubber-mounted earthmoving machine has a platform and attachments that are vertically pivotable so that it can swing freely.

Rheogoniometry: Rheological tests to determine various stresses and shear effects on Newtonian and non-Newtonian fluids.

Rheology: Study of deformation and flow of matter, especially non-Newtonian flow of liquids and plastic flow of solids.

Rheometer: A tool for determining the flow properties of solids by measuring the relationship between stress, strain and time.

Ribbon conveyor: A type of screw conveyor with an open space between the shaft and the belt conveyor used for wet or sticky materials that might otherwise accumulate on the spindle.

Ribbon mixer: Device for the mixing of particles, slurries, or pastes of solids

Rig: A tripod, derrick, or drill machine complete with auxiliary and accessory equipment needed to drill.

Rigid body: An idealized extended rigid body whose size and shape are definitely fixed and remain unchanged when forces are applied.

Rigid-body dynamics: Study of the movements of a rigid body under the influence of forces and moments.

Rigidity: The quality or state of resisting change in form.

Rim clutch: A friction-contact clutch that has surface features that exert pressure on the rim from the outside or from the inside.

Ring crusher: A device for crushing solid particles with a rotor having loose crushing rings, held outside by centrifugal force, which crush the raw material when it hits the surrounding shell.

Ring gear: A ring gear in an automotive differential that is driven by a propeller shaft gear and transfers power through the differential to a linear axle.

Ring job: Installation of new piston rings on a piston.

Ring oil: Lubricate the bearing (bearing) by supplying oil to the lubrication point with a ring that rests on the journal and rotates with it and is immersed in a reservoir containing lubricant.

Ring roller mill: A mill in which material is passed by spring-loaded rollers that apply force to the walls of a rotating drum. Also known as a roller mill.

Ripsaw: A weighty tooth power saw used for cutting wood with the grain.

Ritchie's experiment: An experiment using a Leslie cube and a differential air thermometer to demonstrate that the emissivity of a surface is proportional to its absorbency.

Rittinger's law: The law that the energy required to reduce the size of a solid is directly proportional to the resulting increase in surface area.

Riveting hammer: A hammer used for driving rivets.

Roberts' linkage: A type of approximate rectilinear mechanism that, in the early 19th century, provided a practical means of making straight metal guides for guides in a metal planner.

Robins-Messiter system: Stacking conveyor system in which material is fed onto a conveyor belt and fed to one or two wing conveyors.

Rock channeler: A machine used in a quarry to cut an artificial seam in a mass of stone.

Rock drill: A machine for boring quite short holes in rock for blasting purposes; motive power may be com-

pressed air, steam, or electricity.

Rocker arm: In an internal combustion engine, a lever that pivots about its center and is actuated by a pusher at one end to raise and lower the valve stem at the other end.

Rocker cam: A cam that moves with a rocking motion

Rocking furnace: Horizontal cylindrical melting furnace that rolls back and forth on a gearbox frame.

Rocking valve: An engine valve in which a disk or cylinder turns in its seat to permit fluid flow.

Rod mill: A pulverizer operated by the impact of heavy metal rods.

Rod string: The drill rods are connected to form a connecting link between the core barrel and the bit in the wellbore and the drilling rig on the collar of the wellbore.

Rolamite mechanism: Elemental mechanism consisting of two rollers contained in two parallel planes and bounded by a fixed S-shaped belt under tension.

Roll: Rotational or oscillatory motion of an aircraft or similar body about a longitudinal axis through the body; this is called roll for any degree of such rotation.

Roll acceleration: The angular acceleration of an aircraft or rocket relative to its longitudinal axis or X-axis.

Roll axis: The longitudinal axis of an aircraft, missile, or similar body around which it rolls.

Roll crusher: A crusher having one or two toothed rollers to reduce the material.

Roller bearing: Shaft bearing with parallel or tapered steel rollers enclosed between the outer and inner rings.

Roller cam follower: A follower consisting of a rotatable wheel at the end of the shaft.

Roller chain: A chain drive assembled from roller links and pin links

Roller conveyor: A gravity conveyor that has a track of parallel tubular rollers mounted at a defined slope, usually on rolling bearings, at fixed locations along which packed goods, rigid enough to prevent sagging between the rollers, are moved by gravity or motion.

Roller leveling: Leveling of flat material by passing it through a machine having a number of rolls, the axes of which are offset from the average parallel path by a decreasing amount.

Roller pulverizer: A pulverizer operated by the crushing action of rotating rollers

Roller stamping die: An engraved roller used for stamping designs and other markings on sheet metal.

Rolling: The movement of the body along the surface is combined with the rotational movement of the body, so that the point on the body that is in contact with the surface is instantly at rest.

Rolling contact: The contact between the bodies is such that the relative velocity of the two contacting surfaces at the point of contact is zero.

Rolling-contact bearing: A bearing consisting of rolling elements located between the outer and inner rings.

Rolling friction: A force that opposes the motion of anybody rolling on the surface of another.

Roll mill: A series of rolls operating at different speeds for grinding and crushing.

Roll threading: Threading

a metal billet by rolling it between corrugated circular rolls or between corrugated straight lines.

Rood: A unit of area, equal to 1/4 acre, or 10,890 square feet, or 1011.7141056 square meters.

Roots blower: A compressor in which a pair of hourglass-shaped elements rotate within a housing to deliver large volumes of gas at relatively low pressure increments.

Rope-and-button conveyor: A conveyor consisting of an endless rope or cable with discs or buttons attached at regular intervals.

Rope drive: A system of ropes passing through grooved pulleys or pulleys to transmit power over distances too long for belt drives.

Rossby diagram: Thermodynamic diagram, named after its creator, with mixture ratio on the abscissa and potential temperature on the ordinate; added lines of constant equivalent potential temperature.

Ross feeder: Chute for conveying bulk materials using a screen made of heavy endless chains suspended on the sprocket shaft; the rotation of the shaft causes the materials to slide.

Rotary actuator: A device that converts electrical energy into a controlled rotational force; usually consists of an electric motor, gearbox and limit switches.

Rotary air heater: A regenerative air heater in which heat-transferring members are moved alternately through the gas and air streams.

Rotary annular extractor: Vertical cylindrical body with an inner rotating cylinder; the liquids in contact flow in countercurrent through the annular space between the rotor and the casing; used for liquid-liquid extraction processes.

Rotary atomizer: Hydraulic sprayer with combined pump and nozzle.

Rotary belt cleaner: A row of blades, symmetrically positioned about the axis of rotation, are scratched or bumped against the conveyor belt for cleaning purposes.

Rotary blower: Piston, rotor impeller, air outlet; can be straight, screw, plate or liquid piston type.

Rotary boring: A drilling system in which rock penetration is achieved by rotating a hollow cutter.

Rotary bucket: A drilling device with a diameter of 12 to 96 inches (30 to 244 centimeters), the lower end of which is equipped with cutting teeth used for rotary drilling of shallow holes of large diameter in order to obtain soil samples above the water table.

Rotary compressor: A volumetric machine in which the compression of the liquid is carried out directly by the rotor and without the usual piston, connecting rod and crank mechanism of a reciprocating compressor.

Rotary crane: A crane consisting of a boom pivoted to a fixed or movable structure

Rotary crusher: A device for grinding solid particles in which a high-speed rotating cone on a vertical shaft pushes the solid particles against the surrounding shell.

Rotary cutter: A device used to cut hard or fibrous materials by shearing action between two sets of blades, one mounted on a rotating holder and the other stationary on the surrounding casing.

Rotary drill: Any of a variety of drilling rigs that rotate a rigid tubular rod string to which a drill bit is attached, such as an oil drilling device.

Rotary drilling: The act or process of drilling a well with a rotary drilling rig, for example, when drilling an oil well.

Rotary dryer: The cylindrical kiln is slightly inclined to the horizontal and rotates on suitable bearings; moisture is removed by rising hot gases.

Rotary engine: A positive displacement engine (such as a steam engine or an internal combustion engine) in which the thermodynamic cycle is performed in a mechanism that is fully rotating and without the more conventional reciprocating piston, connecting rods, and crankshaft components.

Rotary feeder: A device in which a rotating element or vane ejects powder or granules at a predetermined speed.

Rotary furnace: A rotary furnace is a barrel-shaped instrument that is rotated around its axis when performing heat treatment. These instruments are tilted slightly to allow the sample under heat treatment to be passed from one end of the barrel to the other.

Rotary-percussive drill: A drilling machine that works like a rotary machine by repeatedly hitting the chisel.

Rotary pump: A displacement pump that delivers a steady flow by the action of two members in rotational contact.

Rotary roughening: A metal preparation technique in which the surface of the workpiece is roughened with a cutting tool.

Rotary shear: Sheet metal cutting machine with two rotating disc cutters mounted on parallel shafts and synchronously controlled.

Rotary shot drill: A rotary drill used to drill blastholes.

Rotary swager : A machine for reducing the diameter or wall thickness of a bar or pipe by striking the surface of a workpiece with a hammer on a mandrel.

Rotary table: Attachment of a milling machine, consisting of a round table with T-shaped slots, rotated by a flywheel that drives a worm-and-worm gear.

Rotary valve: : A valve for admitting or discharging working fluid into or out of an engine cylinder, where the valve element is an orifice piston that rotates around its axis.

Rotating coordinate system: A coordinate system whose axes rotate in an inertial coordinate system.

Rotation: Also known as rotational motion. Motion of a rigid body in which either one point is fixed, or all the points on a straight line are fixed.

Rotational energy: The kinetic energy of a rigid body due to rotation.

Rotational impedance: A complex quantity equal to the vector representing the variable torque acting on the system divided by the vector representing the resulting angular velocity in the direction of the torque at its point of application.

Rotational reactance: Imaginary part of the rotational impedance. Also known as mechanical rotational reactance.

Rotational resistance: The real part of the rotational impedance; he is responsible for dissipating energy. Also known as mechanical resistance to rotation.

Rotational stability: A property of a body for which

a small angular displacement creates a restoring moment that tends to return the body to its original position.

Rotational strain: Strain in which the orientation of the axes of strain is changed.

Rotational traverse: The maximum angle through which the body can rotate, while one of its points remains stationary on the axis or in the center.

Rotation coefficients: Factors used in calculating the effect of Earth's rotation on range and deflection; they are published only in comparatively long range shooting tables.

Rotator: Rotating body. coarse grinding: preliminary grinding without finishing.

Rough machining: Preliminary machining without regard to finish.

Roughness width cut off: The maximum width of the surface irregularities taken into account when measuring the roughness height.

Rough turning: The removal of excess stock from a workpiece as rapidly and efficiently as possible.

Rout: To gouge out, make a furrow, or otherwise machine

a wood member.

Router: Machine with fast rotating vertical spindle and milling cutter for making grooves, grooves and similar grooves.

Routh's procedure: The procedure for modifying the Lagrangian of the system in such a way that the modified function satisfies the modified form of the Lagrange equations, in which the ignored coordinates are excluded.

Runge vector: A vector that describes some of the invariable features of the nonrelativistic interaction of two bodies obeying the inverse square law, both in classical and quantum mechanics; its constancy is a reflection of the inherent symmetry of inverse square interactions.

Running gear: Means used to support a truck and its load, as well as to ensure rolling contact with the working surface.

Run-time data: Information obtained from sensors during normal machine operation and used to improve its performance.

Rupture disk device: Non-reclosing pressure relief device that reduces static inlet pressure to the system due to disc rupture.

Rzeppa joint: A special application of the Bendix Weiss universal joint, in which the four large balls are transmission elements and the central ball acts as a spacer; it transmits constant angular velocity through a single universal joint.

Sabathé's cycle: The cycle of an internal combustion engine in which part of the combustion is explosive and part is at constant pressure.

Saddle type turret lathe: Turret lathe, designed without a slide, with a turret mounted directly on a support (saddle) that slides over the lathe bed.

Safety hoist: Lifting device that does not continue to operate when de-energized.

Safety stop: Attachment of the lifting device, a device with which it is possible to prevent the fall of the load.

Safety valve: A spring-loaded pressure valve that allows steam to escape from the boiler at a pressure slightly above the safe operating level of the boiler; established by law for all boilers.

Sander: An electric machine used to sand the surface of wood, metal, or other material.

Sand hill analogy: Formal identity between the differential equation and the boundary conditions for the torsional stress function of a perfectly plastic prismatic rod and for the surface height of a granular material such as dry sand that has a constant angle of rest. Also known as sand heap analog.

Sand mill: A variant of a ball-type grinding mill in which grains of sand serve as grinding balls.

Sand pump: A pump, usually of a centrifugal type, capable of pumping fluids containing sand and gravel without clogging or excessive wear; It is used to extract drilling mud and cuttings from the well.

Sand reel: A drum, driven by a band wheel, for raising or lowering the sand pump or bailer during drilling operations. Also known as coring reel.

Sand slinger: A machine which delivers sand to and fills molds at high speed by centrifugal force.

Sand wheel: A wheel equipped with steel buckets around the circumference to lift sand or slurry from the sump to a higher level.

Sargent cycle: An ideal thermodynamic cycle consisting of four reversible processes: adiabatic compression, constant volume heating, adiabatic expansion, and isobaric cooling.

Saturated vapor: Steam, the temperature of which is

equal to the boiling point at the pressure existing on it.

Saturation specific humidity: Thermodynamic function of state; the value of the specific humidity of saturated air at a given temperature and pressure.

Saturation vapor pressure: The vapor pressure of a thermodynamic system at a given temperature at which the vapor of a substance is in equilibrium with a flat surface of a pure liquid or solid phase of this substance.

Saunders air-lift pump: A device for raising water from a well by introducing compressed air below the water level in the well.

Savonius rotor: The rotor consists of two offset semi-cylindrical elements rotating around a vertical axis.

Savonius windmill: A windmill consisting of two semi-cylindrical offset cups rotating around a vertical axis.

Saw gumming: Grinding away the punch marks in the spaces between the teeth in saw manufacture.

Sawmill: A machine for cutting logs with a saw or a series of saws.

Saw tooth crusher: A solids crusher in which the feed is crushed between two saw-tooth shafts rotating at different speeds.

Scavenging: Removal of exhaust gases from the internal combustion engine cylinder and replacement with fresh charge or air.

Scheffel engine: A type of multi-rotor motor that uses nine roughly identical rotors that rotate in the same clockwise direction.

Schleiermacher's method: A method for determining the thermal conductivity of a gas, in which gas is placed in a cylinder with an electrically heated wire along its axis, and the electrical energy supplied to the wire is measured, as well as the temperature of the wire and the cylinder.

Schneider recoil system: A recoil system for artillery, employing the hydro pneumatic principle without a floating piston.

Schuler pendulum: Any apparatus that swings under the influence of gravity with a natural period of 84.4 minutes, that is, with the same period as a hypothetical simple pendulum, the length of which is equal to the radius of the Earth; the pendulum arm remains vertical despite any movement of its axis, and therefore the device is useful in navigation.

scissor jack: Jack driven by a horizontal screw; jack levers are parallelograms, the horizontal diagonals of which are lengthened or shortened by a screw.

Scotch boiler: Fire-tube boiler with one or more cylindrical internal furnaces surrounded by a boiler casing having five pipes in the upper part; heat is transferred to the water partly in the furnace zone and partly when hot gases pass through the pipes.

Scotch yoke: Type of four-link hitch; it is used to transform stable rotation into simple harmonic movement.

Scraper conveyor: A type of scraper conveyor in which an element (chain and scraper) is supported on a chute to move materials.

Scraper hoist: A drum hoist that operates the scraper of a scraper loader.

Scraper loader: A machine used to load coal or rock by pulling a bucket through material to an apron or ramp, from where the load is unloaded onto a vehicle or conveyor.

Scraper ring: A piston ring that scrapes oil off the cylinder wall to prevent burns.

Screw compressor: A gas compressor with a rotary element in which compression is carried out between two mutually rotating screws rotating in opposite directions.

Screw conveyor: A conveyor consisting of a screw that rotates on a single shaft inside a stationary chute or housing and can move bulk material in a horizontal, inclined or vertical plane. Also known as screw conveyor; screw conveyor; screw conveyor.

Screw displacement: Rotation of a rigid body around an axis, accompanied by the movement of the body along the same axis.

Screw elevator: A type of screw conveyor for vertical delivery of pulverized materials.

Screw feed: A system or combination of gears, ratchets and friction devices in the rotary head of a diamond drill that controls the speed at which the bit penetrates the rock.

Screw feeder: A mechanism for handling bulk materials (crushed or granular solids), in which a rotating screw propeller moves the material forward, to and into the process unit.

Screw machine: A lathe for the production of relatively small, turned metal parts in large quantities.

Screw press: A press having the slide operated by a screw mechanism.

Screw propeller: Marine and aircraft propeller, consisting of a streamlined hub attached externally to a rotating engine shaft, on which two to six blades are mounted; the blades form helical surfaces in such a way as to move along the axis around which they rotate.

Screw pump: A pump that lifts water using screw impellers in the pump casing.

Screw stock: Free machining bar, rod, or wire.

Secondary air: Combustion air is introduced above the burner flame to improve combustion efficiency.

Secondary creep: Change in the shape of a substance at a minimum and almost constant differential stress with a constant dependence of deformation on time. Also known as steady-state creep.

Secondary crusher: Any of a group of crushing and shredding machines used after primary processing to further reduce the particle size of shale or other rock.

Secondary grinding: A further grinding of material previously reduced to sand size.

Secondary stress: Self-limiting normal or shear stress that is caused by the constraint of a structure and is expected to cause minor deformations that will not result in structural failure.

Second law of thermodynamics: The general statement that there is a preferred direction for any process; There are many equivalent formulations of the law, the most famous of which are Clausius and Kelvin.

Second-order transition: A change in state in which the free energy of a substance and its first derivatives are continuous functions of temperature and pressure or other relevant variables.

Section modulus: The ratio of the moment of inertia of the cross-section of a beam subjected to bending to the greatest distance of a beam element from the neutral axis.

Selective transmission: A gear transmission with a single lever for changing from one gear ratio to another; used in automotive vehicles.

Self centering chuck: Drill chuck that, when closed, automatically positions the drill rod in the center of the drive rod of the diamond drill articulated head.

Self propelled: Pertaining to a vehicle given motion by means of a self-contained motor.

Self starter: An attachment for automatically starting an internal combustion engine.

Sellers hob: A hob that rotates the centers of the lathe and the work is fed to it by the lathe carriage.

Selwood engine: A rotating block engine in which two curved pistons 180 opposed to each other rotate along toroidal tracks, causing the entire cylinder block to rotate.

Semiautomatic transmission: An automotive transmission that assists the driver in shifting from one gear to another.

Semiclosed cycle gas turbine: A heat engine in which part of the expanded gas is recirculated.

Semi diesel engine: An internal combustion engine, a type of diesel engine that uses heavy oil as fuel, but uses a lower compression pressure and sprays it under pressure onto a hot (uncooled) surface or spot, or ignites them by precombustion or over-compression of the part. charge in a separate cell or uncooled part of the combustion chamber.

Semi floating axle: A support element in automobiles that receives torque and wheel loads at its outer end.

Sensible heat: Heat absorbed or released by a substance when the temperature changes, not accompanied by a change in state.

Sensible heat factor: The ratio of the physical heat of the room to the total amount of heat in the room; used for air conditioning calculations.

Sensible heat flow: Heat given off or absorbed by the body when cooled or heated as a result of the body's ability to retain heat; eliminates latent heat of melting and evaporation.

Servoarm attachment: A device that enhances the maximum distance over which the manipulator of a simple robot can travel.

Servo brake: A brake in which the movement of a vehicle is used to increase pressure on one of the pads. A brake in which the force applied by the operator is increased by a mechanical drive.

Servo valve: A transducer in which the low energy signal controls the flow of the high energy fluid so that the flow is proportional to the signal.

Setback force: The backward inertial force generated by the forward acceleration of a projectile or rocket during the launch phase; forces are directly proportional to the acceleration and mass of the accelerated parts.

Set forward: The relative translational movement of components that occurs in a projectile, missile or bomb in flight upon impact; the effect is caused by inertia and is the opposite of the direction of retreat.

Set forward force: The translational force of inertia, which is created by the deceleration of a projectile, missile or bomb upon impact; forces are directly proportional to the deceleration and the mass of the decelerated parts. Also known as impact force.

Set forward point: The point on the expected course of the target that the target is predicted to arrive at the end of the flight time.

Set pressure: The inlet pressure at which the relief valve begins to open as required by the codes or standards applicable to the protected pressure vessel.

Settling velocity: The speed achieved by a particle falling through a liquid depends on its size and shape, as well as on the difference between its specific gravity and the specific gravity of the settling medium; used to sort particles by grain size.

Shaft: A cylindrical piece of metal used to hold rotating parts of a machine, such as pulleys and gears, to transmit power or motion.

Shaft hopper: Hopper that feeds shafts or pipes to grinders, threading machines, screwdrivers and pipe benders.

Shaft horsepower: The power output of a motor, motor, or other prime mover; or the power input of a compressor or pump.

Shaking screen: The screen used to divide the material into the desired dimensions;

has an eccentric drive or an unbalanced rotating weight that causes shaking.

Shaper: A machine for cutting flat profile surfaces by reciprocating a single-point tool along the work piece.

Shaping dies: A set of dies for bending, pressing or otherwise giving the material the desired shape.

Shattering: Failure to very uneven corner blocks of very hard material subjected to severe stress.

Shaving: Cutting off a thin layer from the surface of a work piece. Trimming uneven edges from stampings, forgings, and tubing.

Shearing die: A die with a punch for shearing the work from the stock.

Shearing forces: Two forces that are equal in magnitude, opposite in direction, and act along two distinct parallel lines.

Shearing machine: A machine for cutting cloth or bars, sheets, or plates of metal or other material.

Shearing punch: A punch that cuts material by shearing it, with minimal crushing effect.

Shearing strain: Distortion resulting from movement of material on opposite sides of a plane in opposite directions parallel to the plane.

Shearing stress: The stress at which a material on one side of a surface presses against a material on the other side of a surface with a force parallel to the surface. Also known as shear stress; shear stress.

Shear plane: A confined zone along which fracture occurs in metal cutting.

Shear spinning: A sheet metal forming process that forms parts with axial symmetry about a mandrel using a tool or roller, in which the deformation is performed by a roller such that the diameter of the original blank does not change, but the thickness of the part is reduced by an amount depending on the angle of the mandrel.

Shear strain: Also known as shift. Deformation of a solid, in which a plane in the body is displaced parallel to itself relative to parallel planes in the body; quantitatively, this is the displacement of any plane relative to the second plane, divided by the perpendicular distance between the planes.

Shear strength: The maximum shear stress which a material can withstand without rupture.

Shear wave: A wave that forces an element of an elastic medium to change its shape without changing its volume. Also known as rotational wave.

Shell pump: Simple pump for removing wet sand or dirt; consists of a hollow cylinder with a ball valve or valve at the bottom.

Shimmy: Excessive vibration of the front wheels of wheeled vehicles, causing jerking of the steering wheel.

Shock: Impulsive or short-term movement or force lasting from thousandths to tenths of a second, capable of causing mechanical resonances; e.g. explosion caused by explosives.

Shock absorber: A spring, a control lever, or a combination thereof, designed to minimize the acceleration of the mass of a mechanism or part of it in relation to its frame or support.

Shock isolation: The application of isolators to alleviate the effects of shock on a mechanical device or system.

Shock mount: A mount used with sensitive equip-

ment to reduce or prevent the transmission of shock movements to the equipment.

Shoe: A metal block used as a form or support in various bending operations.

Shot feed: A device for feeding chilled steel shot at a uniform rate and in appropriate quantities into a circulating fluid flowing downward through a rod or pipe connected to the core barrel and shotgun chisel.

Shovel loader: A wheel loader with a bucket pivotally attached to the chassis that picks up the bulk material, lifts it and unloads it behind the machine.

Shut height: The distance in the press between the bottom of the ram and the top of the bed, indicating the maximum die height that can be placed.

Shutoff head: The pressure generated by a centrifugal or axial pump at zero flow in the system.

Shuttle: Reciprocating motion of a machine that continues to move in one direction.

Shuttle conveyor: Any conveyor in a stand-alone design that moves along a defined path parallel to the material flow.

Side direction: In stress analysis, the direction is perpendicular to the object's symmetry plane.

Side milling: Milling with a sidemilling cutter to machine one vertical surface.

Side rod: A rod connecting the connecting rods of two adjacent driving wheels on one side of the locomotive; distributes power from the main thrust to the drive wheels. One of the rods connecting the piston rod crossheads and engine side arms to the side arms.

Sigma function: The property of a mixture of air and water vapor, equal to the difference between the enthalpy and the product of specific humidity and enthalpy of water (liquid) at thermodynamic wet bulb temperature; it is constant at constant barometric pressure and thermodynamic wet bulb temperature.

Simple pendulum: A device consisting of a small massive body suspended from an inextensible object of low mass on a fixed horizontal axis around which the body and suspension can rotate freely.

Simplex pump: A pump with only one steam cylinder and one water cylinder.

Simultaneity: Two events have simultaneity with respect to the observer if they occur at the same time according to the clock set with respect to the observer.

Single acting: Acting in one direction only, as a single-acting plunger, or a single acting engine (admitting the working fluid on one side of the piston only).

Single-action press: A press having a single slide.

Single-block brake: Friction brake, consisting of a short block fitted to the contour of the wheel or drum and pressed to the surface using a lever on the support; used on railway cars.

Single-degree-of-freedom gyro: A gyroscope whose reference axis of rotation can only rotate freely around one of the orthogonal axes, such as an input or output axis.

Single-piece milling: A milling method whereby one part is held and milled in one machine cycle.

Single-stage compressor: A machine that fully compresses a gas or vapor from suction to discharge conditions without any sequential plurality of elements such as cylinders or rotors.

Single-stage pump: A pump in which the head is developed by a single impeller.

Six-axis system: A robot that has six degrees of freedom, three rectangular and three rotational.

Size reduction: Breaking up large pieces of coal, ore or stone with a primary crusher or small pieces by grinding equipment.

Skip hoist: A basket, bucket, or open car accumulated vertically or on an incline on wheels, rails, or shafts and hoisted by a cable; used to raise materials.

Skip trajectory: Trajectory consisting of ballistic phases alternating with missing phases; one of the main trajectories for the non-powered section of the flight of the descent vehicle or spacecraft returning to the Earth's atmosphere.

Skiving: Removal of material in thin layers or chips with high shear or cutting tool slip.

Slabbing cutter: A face-milling cutter used to make wide, rough cuts.

Slackline cableway: A machine widely used in sand and gravel factories that uses an open-ended dragline bucket suspended from a holder that runs on a track cable that can dig, lift and move materials in one continuous operation.

Slat conveyor: A conveyor consisting of horizontal slats on an endless chain.

Sleeve bearing: A machine bearing in which the shaft turns and is lubricated by a sleeve.

Sleeve valve: An intake and exhaust valve in an internal combustion engine, consisting of one or two hollow bushings that fit around the interior of the cylinder and move with the piston so that their holes align with the intake and exhaust holes in the cylinder at respective stages in the cylinder. cycle.

Slide: The main reciprocating element of a mechanical press guided in the press frame to which a punch or upper die is attached.

Slider coupling: Device for connecting shafts with lateral displacement. Also known as dual clutch; Oldham clutch.

Slide rest: An adjustable slide for holding a cutting tool, as on an engine lathe.

Slide valve: Sliding mechanism to close and open fluid inlets as in some steam engines.

Sliding block linkage: A mechanism in which a crank and a sliding block are used to convert rotary motion into translational motion or vice versa.

Sliding chain conveyor: Transport machine for handling boxes, cans, pipes or similar products on simple or modified links of a set of parallel chains.

Sliding friction: Rubbing of bodies in sliding contact.

Sliding-gear transmission: A transmission system utilizing a pair of sliding gears

Sliding pair: Two adjacent links, one of which is forced to move along a certain path relative to the other; the bottom, or closed, pair is completely limited by the design of the links of the pair.

Sliding vector: A vector whose direction and line of attachment are set, but the attachment point is not set.

Slip friction clutch: A friction clutch designed to slip when too much power is applied to it.

Slipper brake: A plate placed against a moving part to slow or stop it. A plate applied to the wheel of a vehicle or to the track roadway to slow or stop the vehi-

cle.

Slip ratio: For a propeller propeller, actual forward propulsion is related to theoretical propulsion as determined by pitch and rotation.

Slitter: Synchronized type of rotary knife feeder; It is used to precisely cut sheet material such as metal, rubber, plastic or paper into strips.

Slitting: Passing sheet or strip material (metal, plastic, paper or fabric) through rotating knives.

Slope conveyor: A troughed belt conveyor used for transporting material on steep grades.

Slope of fall: Ratio between the drop of a projectile and its horizontal movement; tangent of the angle of fall.

Slotter: A machine tool used for making a mortise or shaping the sides of an aperture.

Slotting: Cutting a mortise or a similar narrow aperture in a material using a machine with a vertically reciprocating tool.

Slotting machine: A vertical reciprocating planer used for grooving and forming the sides of openings.

Slug: A unit of mass in British

gravity equal to the mass that accelerates 1 foot per second per second when acted upon by a force of 1 pound; equals approximately 32.1740 pounds of mass or 14.5939 kilograms.

Small-scale hydropower: Electricity production using hydraulic turbines, the installed capacity of which ranges from 5 kilowatts to 5 megawatts.

Smith-McIntyre sampler: A device for sampling bottom sediments from the ocean floor; digging and lifting mechanisms are independent: the digging bucket is pressed into the sediment before lifting occurs.

Smoke box: A chamber outside the boiler for capturing unburned combustion products.

Smoothing mill: A revolving stone wheel used to cut and bevel glass or stone.

Snagging: Removing surplus metal or large surface defects by using a grinding wheel.

Snow blower: A machine that removes snow from a road surface or sidewalk using a screw knife that pushes the snow into the machine and throws it out a distance.

Snowplow: A device for clearing away snow, as from a road or railway track.

Snubber: A mechanical device consisting essentially of a drum, a spring and a friction band connected between an axle and a frame to retard the recoil of the spring and reduce shaking.

Solar engine: An engine that converts the thermal energy of the sun into electrical, mechanical or cold energy; can be used as a method of propelling a spacecraft, either directly by pressing photons on huge solar sails, or indirectly from solar panels, or from a reflector-boiler combination used to heat a liquid.

Solar heating: Converting solar radiation into heat for technological, heating and kitchen purposes.

Solar pond: A type of non-focusing solar collector consisting of a salt water pool heated by the sun; it is used either directly as a heat source or as an energy source for an electric generator. Also known as a salt gradient sun pond.

Solar power: The conversion of the energy of the sun's radiation to useful work.

Solenoid brake: A device that slows down or stops rotational movement due to the magnetic resistance of a solenoid.

Solenoid valve: A valve actuated by a solenoid, for controlling the flow of gases or liquids in pipes.

Solid box: A solid, unadjustable ring bearing lined with babbitt metal, used on light machinery.

Solid coupling: A flange or compression sleeve used to connect two shafts to create a permanent connection and is usually designed to transmit the full load capacity of the shaft; robust coupling lacks flexibility.

Solid injection system: A fuel injection system for a diesel engine in which a pump pumps fuel through a fuel line and a spray nozzle into the combustion chamber.

Solid shafting: A solid round bar that supports a roller and wheel of a machine.

Sonic drilling: The process of cutting or shaping materials with an abrasive slurry driven by a reciprocating tool attached to an audio-frequency electromechanical transducer and vibrating at sonic frequency.

Sonic sifter: A high speed vibrating apparatus used in particle size analysis.

Space centrode: The path traced by the instantaneous center of a rotating body relative to the inertial frame of reference.

Space cone: A cone in space swept away by the instantaneous axis of a rigid body during Poinsot's motion. Also known as herpolod cone.

Spark-ignition engine: An internal combustion engine in which an electrical discharge ignites an explosive mixture of fuel and air.

Spark knock: Knock in an internal combustion engine precedes the piston reaching top dead center.

Spark lead: The amount by which the spark precedes the arrival of the piston at its top (compression) dead center in the cylinder of an internal combustion engine.

Spatial linkage: A linkage that involves motion in all three dimensions.

Specific energy: The internal energy of a substance per unit mass.

Specific fuel consumption: The weighted fuel consumption required to generate a unit of power or thrust, such as pounds per hour. Abbreviated as SFC. Also known as specific fuel consumption.

Specific gravity: The ratio of the density of a material to the density of some standard material, such as water at a specified temperature.

Specific heat: The ratio of the amount of heat required to raise the temperature of the mass of the material by 1 degree to the amount of heat required to raise the same mass of the reference substance, usually water, the temperature by 1 degree; both measurements are taken at a reference temperature, usually at constant pressure or constant volume.

Specific volume: The volume of a substance per unit mass; it is the reciprocal of the density.

Specific weight: The weight per unit volume of a substance.

Spectral emissivity: The ratio of the radiation emitted by a surface at a specific wavelength to the radiation emitted by a perfect blackbody emitter at the same wavelength and temperature.

Speed: The rate of change in body position without regard to direction; in other words, the magnitude of the velocity vector.

Speed cone: Tapered pulley or pulley consisting of a series of pulleys of increasing diameter forming a stepped taper.

Speed lathe: A lightweight pulley-driven lathe, usually without a carriage or rear gears, used for work in which the tool is operated by hand.

Speed pay load trade off: The relationship between the maximum speed with which a machine can move a work piece and the maximum weight of the work piece.

Speed reducer: A series of gears placed between the motor and the mechanisms it will drive to reduce the rate of power transfer.

Speed reliability trade off: The relationship between the maximum speed at which the machine can move the workpiece and the reliability with which the machine can be achieved to some degree of satisfaction.

Spherical pendulum: A simple pendulum mounted on an axis so that its movement is not limited to a plane; the bob moves on a spherical surface.

Spherical stress: The part of the total stress corresponding to the isotropic hydrostatic pressure; its stress tensor is the unit tensor multiplied by one third of the trace of the total stress tensor.

Spin: Rotation of a body about its axis.

Spin compensation: Overcoming or reducing the effect of the rotation of the projectile with a decrease in the penetrating power of the jet in cumulative ammunition.

Spin decelerating moment: A pair around the axis of the projectile, reducing rotation.

Spinning machine: A machine that winds insulation on electric wire. A machine that shapes metal hollow ware.

Spiral gear: A helical gear that transmits power from one shaft to another, nonparallel shaft.

Spiral jaw clutch: A modification of the square-jaw clutch permitting gradual meshing of the mating faces, which have a helical section.

Spline broach: A broach for cutting straight-sided splines, or multiple keyways in holes.

Split-ring piston pack-

ing: A metal ring mounted on a piston to prevent leakage along the cylinder wall.

Spongy: The property of a robot that has an end effect or has a high compliance such that a small force applied to it results in a large movement.

Spontaneous process: A thermodynamic process that occurs without the use of external factors due to the internal properties of the system.

Spool: The drum of a hoist. The movable part of a slide-type hydraulic valve.

Spool type roller conveyor: A type of roller conveyor in which the rollers are tapered or tapered with a larger diameter at the ends of the roller than at the center.

Spot drilling: Drilling a small hole or depression in the surface of the material to serve as a centering guide for subsequent machining operations.

Spot facing: A finished circular surface around the top of the hole for mounting a bolt head or washer, or for flush mounting of mating parts.

Sprag clutch: A clutch designed to transmit power in one direction only.

Spray chamber: A compartment in an air conditioner where humidification is conducted.

Spray dryer: A machine for drying an atomized mist by direct contact with hot gases.

Spray gun: An apparatus shaped like a gun which delivers an atomized mist of liquid.

Spray nozzle: A device in which a liquid is subdivided to form a stream (mist) of small drops.

Spreader stoker: Coal firing system in which mechanical feeders and distributors form a thin layer of fuel on a movable grate, periodic scrubbing grate or reciprocating continuous grating.

Spreading coefficient: The work performed when one liquid spreads per unit area of another is equal to the surface tension of the stationary liquid minus the surface tension of the spreading liquid minus the interfacial tension between the fluids.

Sprengel pump: An air pump that exhausts by trapping gases between drops of mercury in a tube.

Spring coupling: A flexible coupling with resilient parts.

Spring hammer: Hammer with a mechanical drive, driven by compressed spring or compressed air.

Spring loaded regulator: A pressure-regulator valve for pressure vessels or flow systems; the regulator is preloaded by a calibrated spring to open (or close) at the upper (or lower) limit of a preset pressure range.

Spring modulus: Additional force required to deflect the spring by an additional unit of distance; if a certain spring has an elastic modulus of 100 newtons per centimeter, a 100 newton weight will compress it by 1 centimeter, a 200 newton weight by 2 centimeters, and so on.

Spring pin: An iron rod that is installed between the spring and the axle of the locomotive and which maintains an adjustable pressure on the axle.

Sprocket chain: A continuous chain that meshes with the teeth of a sprocket and thus can transfer mechanical power from one sprocket to another.

Sprung axle: A supporting member for carrying the rear wheels of an automobile

Sprung weight: The weight of the vehicle that is supported by the springs, including the frame, radiator, engine, clutch, transmission, body, cargo, etc.

Square: Denotes a unit of area; if x is a unit of length, a square x is the area of a square whose sides have a length of 1x; for example, a square meter

Square engine: An engine in which the stroke is equal to the cylinder bore.

Square-jaw clutch: A type of positive clutch consisting of two or more jaws of square section which mesh together when they are aligned.

Squaring shear: A machine tool consisting of one fixed cutting blade and the other mounted on a reciprocating traverse; used for cutting sheet metal or sheet.

Squeeze roll: A roller designed to apply pressure to material passing between it and a similar roller.

Stabilizer bar: In a motor vehicle, a shaft connecting the two lower suspension arms to reduce body roll when the vehicle is turning. Also known as anti-roll bar.

Stack effect: The pressure difference between the confined hot gas in a chimney or stack and the cool outside air surrounding the outlet.

Stacker: A machine for lifting goods onto a platform or pitchfork and placing them in tiers; with manual control, electric or hydraulic mechanisms.

Stacker-reclaimer: Equipment that transports and stockpiles materials, and recovers and transports materials to recycling facilities.

Stagger-tooth cutter: Side-milling cutter with successive teeth having alternating helix angles.

Stall torque: The amount of torque provided by a motor at close to zero speed.

Standard ballistic conditions: A set of ballistic conditions, arbitrarily adopted as a standard for calculating firing tables.

Standard free-energy increase: An increase in the Gibbs free energy in a chemical reaction when both the reactants and the reaction products are in their standard states.

Standard heat of formation: The heat required to produce one mole of a compound from its elements in their standard state.

Standard trajectory: The path in the air that the projectile is calculated to follow under the given weather conditions, position and materials, including the specific fuse, projectile and propellant being used; Shooting tables are based on standard trajectories.

State of strain: A complete description, including six components of deformation, deformation in a uniformly deformed volume.

State of stress: A complete description, including six stress components, of a uniformly stressed volume.

Statically admissible loads: Any set of external loads and internal forces that fulfills the conditions necessary to maintain the balance of a mechanical system.

Static friction: A force that resists the onset of a sliding motion of one body over another with which it is in contact.

Static load: A non-varying load; the basal pressure exerted by the weight of a mass at rest, such as the load imposed on a drill bit by the weight of the drill-stem

equipment or the pressure exerted on the rocks around an underground opening by the weight of the superimposed rocks. Also known as dead load.

Static moment: A scalar quantity (for example, area or mass) multiplied by the perpendicular distance from the point associated with the quantity (for example, the centroid of an area or center of mass) to the origin axis. 2. The magnitude of a vector (such as a force, momentum, or directional segment of a straight line) multiplied by the length of the perpendicular falling from the line of action of the vector to the control point.

Static reaction: The force exerted on a body by other bodies which are keeping it in equilibrium.

Statics: A branch of mechanics that considers power and power systems, abstracted from matter, and forces acting on bodies in equilibrium.

Stationary engine: A permanently installed motor, for example in a power plant, factory or mine. steady rest: a device that is used to support long thin parts during turning or grinding and allows them to rotate without eccentric movement.

Steady-state conduction: Thermal conductivity, at which the temperature and heat flux at each point do not change over time.

Steady-state vibration: Vibration, at which the speed of each particle in the system is a continuous periodic quantity.

Steam accumulator: A pressure vessel in which water is heated by steam during off-peak periods and regenerated as steam if necessary.

Steam attemperation: The control of the maximum temperature of superheated steam by water injection or submerged cooling.

Steam boiler: A pressurized system in which water is converted to steam by heat transferred from a higher temperature source, usually combustion products from fuel combustion. Also known as a steam generator.

Steam condenser: A device for maintaining a vacuum at the outlet of a steam prime mover by transferring heat to circulating water or air at the lowest ambient temperature.

Steam drive: Any device that uses the energy generated by the pressure of an expanding steam to move a machine or part of it.

Steam dryer: A device for separating liquid from vapor in a steam supply system.

Steam engine: A thermodynamic device for converting heat from steam into work, usually in the form of a positive displacement piston and cylinder mechanism.

Steam hammer: A forging hammer in which the slider is raised, lowered and driven by a steam cylinder.

Steam-heated evaporator: A design that uses condensing steam as a heat source on one side of the heat exchange surface to vaporize liquid on the other side.

Steam heating: A system in which steam was used as a medium for comfort or process heating.

Steam jacket: A casing applied to the cylinders and heads of a steam engine, or other space, to keep the surfaces hot and dry.

Steam-jet cycle: A refrigeration cycle in which water is used as a refrigerant; high speed steam jets create a high vacuum in the evaporator, causing the water to boil at a low temperature while compressing the vaporized steam to the pressure level in the condenser.

Steam-jet ejector: A fluid acceleration vacuum pump or compressor using the high velocity of a steam jet for entrainment.

Steam line: A graph of the boiling point of water as a function of pressure.

Steam locomotive: A railway propulsion power plant using steam, generally in a reciprocating, non-condensing engine.

Steam nozzle: A streamlined flow structure in which heat energy of steam is converted to the kinetic form.

Steam point: The boiling point of pure water, the isotopic composition of which is the same as that of seawater at standard atmospheric pressure; it is assigned a value of 100 ° C on the 1968 International Temperature Scale of Practice.

Steam pump: A pump driven by steam acting on the coupled piston rod and plunger.

Steam reheater: A steam boiler component in which heat is added to the intermediate pressure steam, which has given some of its energy to expand through a high pressure turbine.

Steam roller: A road roller driven by a steam engine.

Steam separator: A device for separating a mixture of liquid and vapor phases of water. Also known as steam purifier.

Steam shovel: A power shovel operated by steam.

Steam superheater: A boiler component in which physical heat is added to the steam after it has evaporated from the liquid phase.

Steam trap: A device which drains and removes condensate automatically from steam lines.

Steam-tube dryer: Rotary dryer with steam-heated tubes running the entire length of the cylinder and rotating with the dryer jacket.

Steam turbine: A prime mover for converting the thermal energy of steam into the work of a rotating shaft, using the principles of hydraulic acceleration in jet and vane mechanisms.

Steam valve: A valve used to regulate the flow of steam

Steel flex coupling: A flexible coupling consisting of two grooved steel hubs, keyed to their respective shafts and connected by a specially hardened alloy steel element called a grid.

Steering arm: A lever that transmits the steering movement from the steering wheel of a motor vehicle to traction.

Steering brake: Means for turning, stopping or holding a tracked vehicle by individually braking the tracks.

Steering gear: A mechanism, including gear and linkage, for controlling the movement of a vehicle or ship.

Steering wheel: A hand wheel for steering the direction of the wheels of an automobile vehicle or rudder of a ship.

Stefan number: A dimensionless number used in the study of radiant heat transfer, equal to the Stefan-Boltzmann constant multiplied by the cube of temperature multiplied by the thickness of the layer divided by the thermal conductivity of the layer. Symbolizes St.

Stem correction: A correction which must be made in reading a thermometer in which part of the stem, and the thermometric fluid within it, is at a temperature which differs from the temperature being measured.

Stem winding: Pertaining to a timepiece that is wound by an internal mechanism turned by an external knob and stem (the winding button of a watch).

Step bearing: Device supporting the lower end of the vertical shaft. Also known as spherical ball bearing.

Step pulley: A series of pulleys of various diameters combined in one concentric block and used to change the speed ratio of the shafts.

Sthène: The force which, when applied to a body whose mass is 1 metric ton, results in an acceleration of 1 meter per second per second; equal to 1000 newtons. Formerly known as funal.

Stiction: Friction that prevents relative movement between two moving parts in their zero position.

Stiffleg derrick: An oil rig consisting of a mast held upright by a fixed tripod on steel or timber supports.

Stiffness: The ratio of a steady force acting on a deformable elastic medium to the resulting displacement.

Stiffness coefficient: The

ratio of the force acting on a linear mechanical system, such as a spring, to its displacement from an equilibrium position.

Stiffness constant: Any of the ratio coefficients in generalized Hooke's law used to express stress components as linear functions of strain components. Also known as elastic constant.

Stiffness matrix: The K matrix used to express the potential energy V of a mechanical system during small displacements from an equilibrium position,

Stigma: The unit of length used primarily in nuclear measurements is 10^{-12} meters. Also known as bicron.

Stirling cycle: Regenerative thermodynamic energy cycle using two isothermal phases and two constant volume phases.

Stirling engine: An engine that is operated by gas expansion at high temperatures; heat for expansion is supplied through the wall of the piston cylinder.

Stodola method: A method for calculating the deflection of a homogeneous or non-uniform beam with free lateral vibrations at a given fre-

quency.

Stoker: A mechanical agent used in a kiln to feed coal, remove debris, control air supply, and mix with combustibles for efficient combustion.

Stone: A unit of mass commonly used in the United Kingdom, equal to 14 pounds or 6.35029318 kilograms.

Straddle milling: Face milling of two parallel vertical surfaces of a part with two side mills simultaneously

Straddle truck: Self-loading industrial dump truck with outriggers that picks up the load before lifting it between the support arms.

Straight-flow turbine: The horizontal axis of a low-head hydraulic turbine, in which the tanks upstream and downstream are connected by a straight tube in which the runners are integrated, and the generator is located directly on the periphery of these runners.

Straight-line mechanism: The connection is so proportional and limited that a point on it describes part of its movement as a straight or almost straight line.

Straight turning: Work turned in a lathe so that the

diameter is constant over the length of the work piece.

Straightway pump: A pump with suction and discharge valves arranged to give a direct flow of fluid.

Strain: The change in the length of an object in some direction per unit of undistorted length in some direction, not necessarily the same; nine possible deformations form a tensor of the second rank.

Strain ellipsoid: Mathematical representation of the deformation of a homogeneous body by deformation that is the same at all points, or uneven stress at a single point. Also known as strain ellipsoid.

Strain energy: The potential energy stored in the body due to elastic deformation is equal to the work that must be done to create this deformation.

Strain rosette: A pattern of intersecting lines on a surface along which linear strains are measured to determine stresses at a point.

Strain tensor: A second-rank tensor whose components are the nine possible strains

Strap hammer: A heavy hammer, driven and driven by a belt drive, in which the head is secured to a belt, usually leather.

Stratified charge engine: Internal combustion engine using a two-layer fuel charge; the rich mixture is close to the spark plug and combustion helps to ignite the lean mixture in the rest of the cylinder.

Strength: The stress at which material ruptures or fails.

Stress: A force per unit area in a solid material that resists separation, compaction or sliding, which tends to be induced by external forces.

Stress amplitude: One half the algebraic difference between the maximum and minimum stress in one fatigue test cycle.

Stress concentration factor: A theoretical factor K_t expressing the ratio of the greatest stress in the region of stress concentration to the corresponding nominal stress.

Stress crack: An external or internal crack in a solid body (metal or plastic) caused by tensile, compressive, or shear forces

Stress difference: The difference between the

greatest and the least of the three principal stresses.

Stress ellipsoid: Mathematical representation of the stress state at a point determined by the minimum, intermediate and maximum stresses and their intensities.

T **Tackle**: Any arrangement of cables and pulleys for mechanical advantage.

Tail pulley: The pulley at the rear of the conveyor belt, opposite to the normal discharge end, can be a drive pulley or an idle pulley.

Tailstock: A part of a lathe that holds the end of the non-molding part and allows it to rotate freely.

Takeup: A tensioning device in a belt conveyor system for taking up slack of loose parts.

Takeup pulley: Adjustable idler to adapt to changing conveyor belt lengths to maintain proper belt tension.

Tandem roller: A steam or gasoline powered road roller in which the weight is divided between rolls of heavy metal of different diameters, one after the other.

Tangential velocity: Instantaneous linear speed of a body moving along a circular path; its direction is tangent to the circular path at the point under consideration.

Tap drill: A drill used to make a hole of a precise size for tapping.

Tape controlled machine: A machine whose movements are automatically controlled using magnetic or punched tape.

Tape drive See tape transport: A device that transfers power from an actuator to a remote mechanism using flexible belts and pulleys.

Taper-rolling bearing: Roller bearing capable of withstanding axial force with tapered rollers and tapered races.

Tappet : A lever or rocker element moved by a cam to tap or touch another part, such as a tappet or a valve system.

Tapping: Forming an internal screw thread in a hole or other part by means of a tap

Tare: The weight of an empty vehicle or container; subtracted from gross weight to ascertain net weight.

Tear strength: The force needed to initiate or to continue tearing a sheet or fabric.

Telescoping valve: Valve with sliding telescopic elements for regulating the flow of water in the piping with minimal impact on the piping.

Telpher: An electric hoist, suspended from and controlled by a wheel cabin, rolls on a single overhead rail or cable.

Telsmith breaker: A type of rotary crusher often used for primary crushing; It consists of a spindle mounted in a long eccentric bushing that rotates to transmit rotary motion to the crushing head, but gives a parallel stroke, that is, the spindle axis describes a cylinder, and not a cone, as in a rotating suspension spindle.

Temperature: An object property that determines the direction of heat flow when an object is in thermal contact with another object.

Temperature-actuated pressure relief valve: A pressure relief valve which work when subjected to increased external or internal temperature.

Temperature bath: A relatively large volume of a homogeneous substance is maintained at isothermally, so that an object in thermal contact with it is maintained at the same temperature.

Temperature color scale: The relationship between the temperature of an incandescent substance and the color of the emitted light.

Temperature gradient: For a given point, a vector whose direction is perpendicular to the isothermal surface at a point, and whose value is equal to the rate of temperature change in this direction.

Temperature scale: Assigning numbers to temperatures in a continuous manner so that the resulting function is single-valued; it is either an empirical temperature scale based on some convenient property of a substance or object, or it measures absolute temperature.

Ten Broecke chart: A graphical plot of heat transfer and temperature differences used to calculate the thermal efficiency of a counter current cool fluid and warm fluid heat exchange system.

Tender: A car attached to a locomotive and carrying supplies of fuel and water.

Tensile modulus: The tangent or secant modulus of elasticity of a material in tension

Tensile strength: The maximum stress that a material subjected to tensile loading can withstand without breaking. Also known as hot power.

Tensile stress: Stress developed by a material bearing a tensile load.

Tensile test: A test in which a specimen is subjected to increasing longitudinal pulling stress until fracture occurs.

Tension: The condition of a string, wire, or rod that is stretched between two points

Tension pulley: A pulley around which an endless cable runs, mounted on a trolley or other movable bearing, so that the slack in the cable can be easily compensated for by gravity.

Terminal unit: In an air conditioning system, a block at the end of an exhaust duct that carries or delivers air to the conditioned space.

Tertiary air: Combustion air added to primary and secondary air.

Theoretical relieving capacity: The capacity of a theoretically perfect nozzle calculated in volumetric or gravimetric units.

Thermal capacitance: The ratio of the entropy added to a body to the resulting rise in temperature.

Thermal compressor: A steam-jet ejector designed to compress steam at pressures above atmospheric.

Thermal conductance: The amount of heat transmitted by a material divided by the difference in temperature of the surfaces of the material.

Thermal conductivity: The thermal conductivity of a material is a measure of its ability to conduct heat. It is commonly denoted by k.

Thermal conductor: A substance with a relatively high thermal conductivity

Thermal coulomb: A unit of entropy equal to 1 joule per kelvin

Thermal drilling: A machining method in which holes are drilled in a workpiece by heat generated from the friction of a rotating tool.

Thermal equilibrium: Property of a system all parts of which have attained a uniform temperature which is the same as that of the system's surroundings.

Thermal farad: A unit of heat capacity equal to the heat capacity of a body, for which an increase in entropy by 1 joule per kelvin leads to an increase in temperature by 1 kelvin.

Thermal flame safe-

guard: A thermocouple located in the pilot flame of the burner; when the pilot flame is extinguished, the emergency circuit is interrupted and the fuel supply is cut off.

Thermal henry : A unit of thermal inductance equal to the product of the temperature difference of 1 kelvin and the time in 1 second, divided by the entropy consumption of 1 watt per kelvin.

Thermal hysteresis: A phenomenon that is sometimes observed in the behavior of any property of the body depending on temperature; it is considered that this happens if the behavior of such a property when the body is heated in a given temperature range differs from when it is cooled in the same temperature range.

Thermal inductance: The product of temperature difference and time divided by entropy flow.

Thermal instrument: Electrical heating device such as a thermocouple or hot wire device.

Thermal ohm: A unit of thermal resistance equal to thermal resistance for which a temperature difference of 1

kelvin creates an entropy flux of 1 watt per kelvin. Also known as Fourier.

Thermal potential difference: The difference between the thermodynamic temperatures of two points.

Thermal resistance: A measure of a body's ability to prevent heat from passing through it, equal to the difference between temperatures on opposite sides of the body divided by the rate of heat flow. Also known as heat resistance.

Thermal resistivity: The reciprocal of the thermal conductivity

Thermal shock: Stress produced in a body or in a material as a result of undergoing a sudden change in temperature.

Thermal stress: Mechanical stress in the body when some or all of its parts are unable to expand or contract in response to temperature changes.

Thermal stress cracking: Cracking or cracking of materials (plastics or metals) due to excessive exposure to elevated temperatures and sudden temperature changes or large temperature changes.

Thermal transpiration: Formation of a pressure gradient in the gas inside the pipe in the presence of a temperature gradient in the gas and when the free path of molecules in the gas makes up a significant part of the pipe diameter. Also known as thermal effusion.

Thermal valve: A valve controlled by an element made of a material whose properties change significantly in response to temperature changes.

Thermodynamic cycle: A procedure or device in which some material undergoes a cyclic process and one form of energy, such as heat at an elevated temperature from the combustion of fuel, is partially converted to another form, such as mechanical shaft energy, the rest is rejected into a lower temperature absorber

Thermodynamic equation of state: An equation that relates the reversible change in the energy of a thermodynamic system to pressure, volume, and temperature.

Thermodynamic equilibrium: Property of a system which is in mechanical, chemical, and thermal equilibrium.

Thermodynamic function of state: Any of the quantities that determine the thermodynamic state of matter in thermodynamic equilibrium; for an ideal gas, pressure, temperature and density are fundamental thermodynamic variables, any two of which, according to the equation of state, are sufficient to determine the state.

Thermodynamic potential: One of several vast quantities that are determined by the instantaneous state of a thermodynamic system, regardless of its previous history, and which are minimal when the system is in thermodynamic equilibrium under certain conditions.

Thermodynamic principles: Laws governing the conversion of energy from one form to another.

Thermodynamic probability: Under certain conditions, the number of equiprobable states in which a substance can exist; the thermodynamic probability is related to the entropy S by the relation S k ln, where k is the Boltzmann constant.

Thermodynamic process: A change in a system is

defined by a passage from an initial to a final state of thermodynamic equilibrium.

Thermodynamic system: A thermodynamic system is a body of matter and/or radiation, confined in space by walls, with defined permeabilities, which separate it from its surroundings.

Thermodynamic temperature scale: Any temperature scale in which the ratio of the temperatures of two reservoirs is equal to the ratio of the amount of heat absorbed from one of them by a heat engine operating according to the Carnot cycle to the amount of heat removed by this engine to the other reservoir.

Thermometric property: A physical property that changes with temperature in a known manner and therefore can be used to measure temperature.

Thermometry: The science and technology of temperature measurement and the setting of temperature measurement standards.

Thermophoresis: The movement of particles in a thermal gradient from high to low temperatures.

Thermosiphon: A closed system of tubes connected to a water cooled engine which permit natural circulation and cooling of the liquid by utilizing the difference in density of the hot and cool portions.

Thetagram: A thermodynamic diagram with coordinates of pressure and temperature, both on a linear scale.

Third law of thermodynamics: The entropy of all perfect crystalline solids is zero at absolute zero temperature.

Thoma cavitation coefficient: The equation for measuring cavitation in a hydraulic turbine installation, relating vapor pressure, barometric pressure, runner setting, tail water, and head.

Thread cutter: A tool used to cut screw threads on a pipe, screw, or bolt

Threading die: A die, which can be a solid, adjustable or adjustable spring, or a self-opening punching head used to extrude threads on a part.

Threading machine: A tool used to cut or form threads inside or outside a cylinder or cone.

Three-body problem: The problem of predicting the motion of three objects obeying Newton's laws of motion and attracting each other according to Newton's law of gravitation.

Throttle valve: A choking device to regulate flow of a liquid, for example, in a pipeline, to an engine or turbine, from a pump or compressor.

Throw: The maximum diameter of the circle moved by a rotary part.

Throwout: In automotive vehicles, a mechanism or set of mechanisms by which the driven and driven clutch discs are separated.

Thrust: The force exerted in any direction by a fluid jet or by a powered screw.

Thrust bearing: A bearing which sustains axial loads and prevents axial movement of a loaded shaft.

Thrust load: A load or pressure parallel to or in the direction of the shaft of a vehicle

Thrust yoke: The part connecting the piston rods of the feed mechanism on a hydraulically driven diamond-drill swivel head to the thrust block, which forms the connecting link between the yoke and the drive rod, by means of which link the longitudinal movements of the feed mechanism are transmitted to the swivel-head drive rod

Tilting dozer: A dozer that can be rotated on the horizontal center pin to cut low on both sides.

Tilting idlers: An arrangement of guide rollers in which the upper part is mounted on vertical arms that pivot on spindles mounted low on the frame of the roller chair.

Tilting mixer: A small-batch mixer consisting of a rotating drum which can be tilted to discharge the contents; used for concrete or mortar.

Time of flight: Elapsed time in seconds from the instant a projectile or other missile leaves a gun or launcher until the instant it strikes or bursts.

Timing: Adjustment in the relative position of the valves and crankshaft of an automobile engine in order to produce the largest effective output of power.

Timing belt pulley: A pulley that is similar to a flat belt pulley without a rim, except that the grooves for the belt teeth are cut into the surface of the pulley parallel to the axis.

Toe-in: The degree (usually

expressed in fractions of an inch) that the front of a vehicle's front wheels are closer together than the rear is measured at the height of the hub with the wheels in the normal straight-ahead position. steering gear.

Toe-out: The outward inclination of the wheels of an automobile at the front on turns due to setting the steering arms at an angle.

Toggle press: A mechanical press in which a toggle mechanism actuates the slide

Tondal: A unit of force equal to the force which will impart an acceleration of 1 foot per second to a mass of 1 long ton; equal to approximately 309.6911 newtons.

Tonne: A unit of mass in the metric system, equal to 1000 kilograms or to approximately 2204.62 pound mass. Also known as metric ton.

Tool changer: In program guarded machines and robotics, a mechanism that allows the use of multiple tools.

Tool dresser: A tool stone grade diamond inset in a metal shank and used to trim or form the face of a grinding wheel.

Tool head: The adaptable tool carrying part of a machine tool.

Tooling: Tools or end effectors with which a robot performs the actual work on a work piece.

Tool offset: The alteration of tool positions in machines to compensate for their wear, finishing, or displacement from an axis.

Tool post: An apparatus to clamp and position a tool holder on a machine tool.

Top: A solid body, one point of which is fixed in an inactive reference frame, and usually this has an axis of prosperity passing through this point. Its velocity is usually studied when it rotates rapidly in the axis of prosperity.

Top dead center: In a reciprocating engine, the dead centre is the position of a piston in which it is either farthest from, or nearest to, the crankshaft. The former is known as Top Dead Centre while the latter is known as Bottom Dead Centre.

Topple: In gyroscopes for marine or aeronautical use, the condition of a sudden upset gyroscope or a gyroscope platform evidenced by a sudden and rapid precession of the spin axis due to large torque disturbances such as the spin axis striking the mechanical stops. Also known as tumble.

Topple axis: Axis of a gyroscope, the horizontal axis, perpendicular to the horizontal spin axis, around which topple occurs. Also known as tumble axis.

Torque arm: In automotive vehicles, an arm to take the torque of the rear axle.

Torque converter: A device for changing the speed of the torque or mechanical advantage between the input and output shafts.

Torque reaction: On a shaft-driven vehicle, the reaction is between the bevel gear and its shaft (which is supported in the rear axle housing) and the bevel ring gear (which is attached to the differential housing), which tends to rotate the axle housing around the axle. instead of rotating only the semi axes.

Torque-winding diagram: Diagram showing how the load on the winding of the winch drum changes and is used to determine the required balancing method; obtained by plotting torque in pounds per foot on the vertical axis versus time or revolutions or depth on the horizontal axis.

Torr: A unit of pressure, equal to 1/760 atmosphere; it differs from 1 millimeter of mercury by less than one part in seven million; ap prox- imately equal to 133.3224 pascals.

Torsiometer: An instrument that measures power transmitted by a rotating shaft; consists of angular scales mounted around the shaft from which twist of the loaded shaft is determined.

Torsion: A twisting deformation of a solid body about an axis in which lines that were initially parallel to the axis become helices.

Torsional angle: The total relative revolution of the ends of a straight cylindrical bar when subjected to a torque.

Torsional compliance: The reciprocal of the torsional rigidity.

Torsional hysteresis: Dependence of the torques in a twisted wire or rod not only on the present torsion of the object but on its previous history of torsion.

Torsional modulus: The ratio of the torsional rigidity of the bar to its length. Also known as torsion modulus.

Torsional pendulum: A device consisting of a disk or other body with a large moment of inertia mounted on one end of a torsionally flexible elastic rod, the other end of which is held stationary; if the disc is twisted and released, it will perform a simple harmonic motion, provided that the torque in the rod is proportional to the angle of twist. Also known as a torsion pendulum.

Torsional rigidity: The ratio of the torque applied around the center axis of the bar at one end of the bar to the resulting twist angle when the other end is held stationary.

Torsional vibration: Periodic movement of the shaft, in which the shaft is twisted around its axis, first in one direction, and then in the other; this movement can be superimposed on a rotary or other movement.

Torsion bar: A spring flexed by twisting about its axis; found in the spring suspension of truck and passenger car wheels, in production machines where space limitations are critical, and in high-speed mechanisms where inertia forces must be minimized.

Torsion damper: A damper used on automobile internal combustion engines to reduce torsional vibration.

Total coincidence : The condition in which all the joints of a robot become locked in position.

Total pressure: The gross load applied on a given surface.

Toughness: The property of a material capable of absorbing energy during plastic deformation; intermediate link between softness and fragility.

Towed load: The weight of a carriage, trailer, or other equipment towed by a prime mover.

Tracer milling: Cutting a duplicate of a three-dimensional form by using a mastic form to direct the tracer controlled cutter.

Traction: Pulling friction of a moving body on the surface on which it moves.

Traction control system: An acceleration sensor control system that, when the drive tire is not in grip, slows down the wheel by braking or reduces engine speed and torque, unless the braking prevents the wheels from spinning.

Tractor: A four-wheel or track tread vehicle used for towing agricultural or construction implements.

Tractor drill: A drill having a crawler mounting to support the feed-guide bar on an extendable arm.

Tractor loader: Tractor equipped with a tipping bucket that can be used to dig and lift soil and rock debris for dumping to truck height. Also known as shovel bulldozer; tractor shovel.

Tramway: An overhead rail, rope, or cable on which wheeled cars run to convey a load.

Transfer case: In a vehicle with more than one driving axle, a housing fitted with gears that distribute the driving power among the axles.

Transfer machine: Equipment that moves parts from one production location in a factory to another.

Transfer matrix method: A method for analyzing the vibrations of complex systems, in which the system is approximated by a finite number of elements connected in a chain way, and matrices are constructed that can be used to determine the configuration and forces acting on one element in terms of forces acting on one element.

Transition: A change of a substance from one of the three states of matter to another.

Transition point: Either the temperature at which a substance passes from one state of aggregation to another (first-order transition), or the temperature of the culmination of a gradual change, for example, the lambda point, or the Curie point (second-order transition). Also known as transition temperature.

Translation: The linear movement of a point in space without any rotation

Translational motion: Motion of a rigid body in such a way that any line which is imagined rigidly attached to the body remains parallel to its original direction.

Transmissibility: A measure of the ability of a system to either amplify or suppress input vibration, equal to the ratio of the system's response amplitude in steady-state forced vibration to the excitation amplitude; the ratio can be expressed in forces, displacements, speeds or accelerations.

Transporter crane: A long lattice girder supported by two lattice towers that can either be fixed or moved along rails at right angles to the girder; a crab moves along the beam with a winch suspended from it.

Transport vehicle: Vehicle primarily intended for personnel and cargo carrying

Trapezoidal excavator: A digging machine which removes earth in a trapezoidal cross-section pattern for canals and ditches.

Travel: The vertical distance of the path of an elevator or escalator as measured from the bottom terminal landing to the top terminal landing.

Traveling block: The movable block, consisting of pulleys, frame, fork and hook, is connected to the load, rises or falls with it in the block-gripping system. Also known as a floating block; running block.

Traveling-grate stoker: Type of furnace stoker; coal is fed by gravity into a hopper located at one end of the movable (movable) grate; When the grate passes under the hopper, it transfers the fresh coal layer to the

furnace.

Tray elevator: A device for lifting drums, barrels, or boxes; a parallel pair of vertical-mounted continuous chains turn over upper and lower drive gears, and spaced trays on the chains cradle and lift the objects to be moved.

Trepanning tool: Cutting tool in the form of a round tube with teeth at the end; The workpiece or tube, or both, rotates and the tube is fed axially into the workpiece, leaving behind a narrow grooved surface in the workpiece.

Tresca criterion: The assumption that plastic deformation of the material begins when the difference between the maximum and minimum principal stresses is equal to twice the shear yield strength.

Triangle of forces: A triangle, two of whose sides represent forces acting on a particle, while the third represents the combined effect of these forces.

Trigger pull: Resistance offered by the trigger of a rifle or other weapon; force which must be exerted to pull the trigger.

Trimmer conveyor: A

self-contained, lightweight, portable conveyor, usually of a belt type, for use in unloading and transporting bulk materials from trucks to indoor storage areas, and for trimming bulk materials in silos or stacks.

Trip hammer: A large power hammer whose head is tripped and falls by cam or lever action.

Triplex chain block: A geared hoist using an epicyclic train

Tripod drill: A reciprocating rock drill mounted on three legs and driven by steam or compressed air; the drill steel is removed and a longer drill inserted about every 2 feet (61 centimeters).

Tripper : A device that snubs a conveyor belt causing the load to be discharged.

Trolley: A wheeled car running on an overhead track, rail, or ropeway. 2. An electric streetcar.

Trolley locomotive: A locomotive operated by electricity drawn from overhead conductors by means of a trolley pole.

Troughed belt conveyor: Belt conveyor with raised edges of the conveyor belt on the carrier portion to form a groove by conforming to the shape of the grooved guide rollers or other bearing surface.

Troughed roller conveyor: A roller conveyor having two rows of rolls set at an angle to form a trough over which objects are conveyed.

Troughing idler: A belt idler having two or more rolls arranged to turn up the edges of the belt so as to form the belt into a trough

Troughing rolls: The rolls of a troughing idler that are so mounted on an incline as to elevate each edge of the belt into a trough

Trouton's rule: The rule that, for an unbound liquid, the latent heat of vaporization in calories is about 22 times the normal boiling point on the Kelvin scale.

Troweling machine: A motorized device used to spread concrete by operating orbiting steel trowels on radial arms rotated on a vertical shaft.

Troy system: A system of units of mass used primarily for measuring gold and silver; ounce is the same as in the pharmaceutical system, is 480 grains or 31.1034768 grams.

Truck: Self-propelled wheeled vehicle designed primarily for the transport of goods and heavy equipment; it can be used to tow trailers or other mobile equipment.

Truck crane: A crane carried on the bed of a motortruck

Truck-mounted drill rig: A drilling rig mounted on a lorry or caterpillar tracks

True rake: The angle, measured in degrees, between a plane containing a tooth face and the axial plane through the tooth point in the direction of chip flow.

Truing: Cutting a grinding wheel to make its surface run concentric with the axis.

Tschudi engine: A cat-and-mouse engine in which the pistons, which are sections of a torus, travel around a toroidal cylinder; motion of the pistons is controlled by two cams which bear against rollers attached to the rotors

Tube bank: An array of tubes designed to be used as a heat exchanger.

Tube cleaner : A device equipped with cutters or brushes used to clean tubes in heat transfer equipment.

Tube door: A door in a boiler furnace wall which facilitates the removal or installation of tubes.

Tube mill: A revolving cylinder used for fine pulverization of ore, rock, and other such materials; the material, mixed with water, is fed into the chamber from one end, and passes out the other end as slime.

Tube turbining: Cleaning tubes by passing a power-driven rotary device through them.

Tumbler gears: Idler gears interposed between spindle and stud gears in a lathe gear train; used to reverse rotation of lead screw or feed rod.

Tumbling mill: A grinding and pulverizing machine consisting of a shell or drum rotating on a horizontal axis.

Tunnel carriage: A machine used for rapid tunneling, consisting of a combined drill carriage and a water / air manifold, so that as soon as the carriage reaches the bottom hole, drilling can begin without wasting time connecting or waiting for drill steel.

Turbine: A fluid acceleration machine for generating rotary mechanical power from the energy in a stream of fluid.

Turbine propulsion: Propulsion of a vehicle or vessel by means of a steam or gas turbine.

Turbining: The removal of scale or other foreign material from the internal surface of a metallic cylinder.

Turbo blower: A centrifugal or axial flow compressor

Turbo pump: A pump that is powered by a turbine.

Turbo shaft: A gas turbine engine that is similar to a turboprop but operates through a transmission system to power a device such as a helicopter rotor or pump.

Turbo supercharger: A gas turbine driven centrifugal air compressor commonly used to pressurize the intake system in a reciprocating internal combustion engine.

Turning : It is one of the operation of a Lathe. Shaping a member on a lathe.

Turning-block linkage: A type of sliding block mechanical link in which the short link is secured and the frame can rotate freely.

Turning center: A numerically controlled lathe that sometimes functions together with a robot in boring and other machining work

Turret lathe: A semi-automatic lathe, which differs from a motorized lathe in that the tailstock is replaced by a multi-faceted dividing tool holder or a turret designed to hold multiple tools.

Twin cable ropeway: Elevating cable car with parallel support cables and supporting vehicles running in opposite directions; both rows of carriers pull the same traction cable.

Twin-geared press: A crank press having the drive gears attached to both ends of the crankshaft.

Two-body problem: The problem of predicting the motion of two objects obeying Newton's laws of motion and forces acting on each other according to a certain law, for example, Newton's law of gravitation, taking into account their mass, position and speed at a certain initial moment of time.

Two-cycle engine: A reciprocating internal combustion engine that requires two piston strokes or one revolution to complete a cycle.

Two-degrees-of-freedom gyro: A gyroscope whose axis of rotation can rotate freely around two orthogonal axes, not counting the axis of rotation.

Two-lip end mill: An end-milling cutter having two cutting edges and straight or helical flutes.

Two-stroke cycle: An internal combustion engine cycle where four process completed in two strokes of the piston.

U

U-bend die: The die with a square or rectangular cross section that offers two edges over which metal can be drawn

Ultimate strength: It is the maximum stress that a material can withstand while being stretched or pulled.

Ultrasonic atomizer: An atomizer in which liquid is fed to, or caused to flow over, a surface which vibrates at an ultrasonic frequency; uniform drops may be formed at low feed rates.

Ultrasonic drilling: A vibration drilling technique that can be used in drilling, cutting, and shaping of hard materials. In this method, ultrasonic vibrations are generated by the compression and extension of a core of electrostrictive or magnetostrictive material in a rapidly alternating electric or magnetic field.

Ultrasonic machining: Ultrasonic machining is a subtractive manufacturing process that removes material from the surface of a part through high frequency, low amplitude vibrations of a tool against the material surface in the presence of fine abrasive particles.

Unavailable energy: That part of the energy which, when an irreversible process takes place, is initially in a form completely available for work and is converted to a form completely unavailable for work.

Underdrive press: A mechanical press having the driving mechanism located within or under the bed.

Underhung crane: An overhead traveling crane in which the end trucks carry the bridge suspended under the rails.

Undershot wheel: A water wheel operated by the impact of flowing water against blades attached around the periphery of the wheel.

Under spin: Property of a projectile having insufficient rate of spin to provide proper stabilization.

Uniflow engine: A uniflow engine is a piston engine where gas flow through the cylinder proceeds in a single unidirectional flow, without reversals between strokes

Uniform circular motion: Circular motion in which the angular velocity remains stable.

Uniform load: A load distributed uniformly over a

portion or over the entire length of a beam; measured in N/m.

Unitary air conditioner: A small self-contained electrical unit enclosing a motor driven refrigeration compressor, evaporative cooling coil, air-cooled condenser, filters, fans, and controls.

Unit heater: A heater consisting of a fan for circulating air over a heat-exchange surface, all covered in a common covering.

Unit strain: For tensile strain, the elongation per unit length. For compressive strain, the shortening per unit length. For shear strain, the change in angle between two lines originally perpendicular to each other.

Unit stress: A stress upon a structure at a certain place, expressed in units of force per unit of cross-sectional area.

Universal dividing head: An accessory fixture on a milling machine that rotates the work piece to specified angles between machining steps.

Universal grinding machine: A grinding machine having a swivel table , headstock, and a wheel head that can be rotated on its base.

Universal joint: A universal joint is a joint or coupling connecting rigid rods whose axes are inclined to each other.

Unloader: A power device for removing bulk materials from railway freight cars or highway trucks; in the case of railway cars, the car structure may aid the unloader;

Unloading conveyor: Any of several types of portable conveyors adapted for unloading bulk materials, packages, or objects from conveyances.

Unsin engine: A type of rotary engine in which the trochoidal rotors of eccentric-rotor engines are replaced with two circular rotors, one of which has a single gear tooth upon which gas pressure acts, and the second rotor has a slot which accepts the gear tooth.

Unsprung axle: A rear axle in an automobile in which the housing carries the right and left rear-axle shafts and the wheels are mounted at the outer end of each shaft

Unsprung weight: The weight of the various parts of a vehicle that are not carried on the springs, such as wheels, axles, and brakes.

Updraft carburetor: For a gasoline engine, a fuel-air mixing device in which both the fuel jet and the air-flow are upward.

Updraft furnace: A furnace in which volumes of air are supplied from under the fuel bed or supply.

Upmilling: Milling a work piece by rotating the cutter against the route of feed of the workpiece.

U-tube manometer: A manometer having two limbs, consisting of a U-shaped glass tube partly filled with a liquid of known specific gravity; when the legs of the manometer are connected to separate sources of pressure, the liquid rises in one leg and drops in the other; the difference between the levels is proportional to the difference in pressures and inversely proportional to the liquid's specific gravity.

V

Vacuum brake: Braking system employed on trains. A form of air brake which operates by maintaining low pressure in the actuating cylinder; braking action is produced by opening one side of the cylinder to the atmosphere.

Vacuum cleaner: An electrically motorized mechanical appliance for the dry removal of dust and loose dirt from rugs, fabrics, and other surfaces.

Vacuum drying: Taking away of liquid from a solid material in a vacuum system; used to lower temperatures needed for evaporation to avoid heat damage to sensitive material.

Vacuum evaporation: Vacuum evaporation is the process of causing the pressure in a liquid-filled container to be reduced below the vapor pressure of the liquid.

Vacuum heating: A two-pipe steam heating arrangement in which a vacuum pump is used to maintain a suction in the return piping, thus creating a positive return flow of air and condensate.

Vacuum pump: A compressor for draining air and non-condensable gases from a space that is to be maintained at sub atmospheric pressure.

Vacuum support: That portion of a rupture disk device which avert deformation of the disk resulting from vacuum or rapid pressure change.

Valve: An apparatus used to regulate the flow of fluids in piping systems and machinery.

Valve follower: A linkage between the cam and the push rod of a valve train.

Valve guide: A path which supports the stem of a poppet valve for maintenance of alignment.

Valve head: A valve head or cylinder head is a part of an internal combustion engine that contains and houses the intake and exhaust valves.

Valve lifter: In an internal combustion engine, device for opening the valve of a cylinder.

Valve stem: A **valve stem** is a self-contained valve which opens to admit gas to a chamber.

Valve train: The valves and valve operating method for the control of fluid flow to and from a piston-cylinder machine, for example, steam, diesel, or gasoline engine.

Vane motor rotary actuator: A type of rotary motor actuator which consists of a rotor with several spring-loaded sliding vanes in an elliptical chamber.

Vapor: A gas at a temperature under the critical temperature, so that it can be liquefied by compression, without lowering the temperature.

Vapor compression cycle: it is a refrigeration cycle, it consists of compressor, condenser, expansion valve and evaporator.

Vapor cycle: It is a thermodynamic cycle, operating as a heat engine or a heat pump, during which the working substance is in, or passes through, the vapor state.

Vaporization coefficient: The vaporization coefficient, αv, is usually defined as the ratio of the actual flow of gaseous decomposition product J to the flow Jmax coming from an effusion cell, in which, it is assumed, decomposition products are in an ideal equilibrium with the reactant.

Vapor pressure: The pressure exerted by a vapor in thermodynamic equilibrium with its condensed phases.

Variable force: A force whose direction, magnitude, or both change with time.

Variable speed drive: A method transmitting motion from one shaft to another that permits the velocity ratio of the shafts to be varied continuously.

Variable volume air system: An air conditioning system in which the volume of air delivered to each controlled zone is varied automatically from a preset minimum to a maximum value depending on the load in each zone.

Varignon's theorem: This theorem states that the moment of a force is the algebraic sum of the moments of its vector components acting at a common point on the line of action of the force.

Variometer: A device for indicating an aircraft's rate of climb or descent. A geomagnetic device for detecting and indicating changes in one of the components of the terrestrial magnetic field vector, usually magnetic declination, the horizontal intensity component, or the vertical intensity component.

V-bend die: A die with a triangular crosssectional opening to offer two edges over which bending is accomplished.

V-bucket carrier: A conveyor consisting of two strands of roller chain separated by V-shaped steel buckets; used for elevating and conveying nonabrasive equipment, such as coal.

Vehicle: A self-propelled wheeled device that transports people or goods on or off roads; automobiles and trucks are examples.

Velocity: The rate at which an object changes its position. it is a vector quantity having direction as well as magnitude. Also known as linear velocity.

Velocity analysis: A graphical method for the determination of the velocities of the parts of a mechanical device, especially those of a plane mechanism with rigid component links.

Velocity ratio: The ratio of the velocity given to the effort or input of a machine to the velocity obtained by the load or output

Ventilator: A mechanical equipment for producing a current of air, as a blowing or exhaust fan.

Vertical band saw: A band saw whose blade operates in the vertical plane, ideal for contour cutting.

Vertical boiler: A fire-tube boiler having vertical tubes between top head and tube sheet, connected to the top of an internal furnace.

Vertical boring mill: A large type of boring machine in which a rotating work piece is fixed firmly to a horizontal table, which resembles a four jaw independent chuck with extra radial T- slots, and the tool has a traverse movement.

Vertical broaching machine: A broaching machine having the broach climbed in the vertical plane.

Vertical conveyor: A materials handling machine planned to move or transport bulk materials or packages upward or downward.

Vertical firing: The discharge of fuel and air perpendicular to the burner in a furnace.

Vertical guide idlers: Idler rollers about 3 inches (8 centimeters) in diameter so placed as to make contact with the edge of the belt conveyor.

Vertical traverse: The angle through which a robot's

arm can sway up and down, typically 30 .

V guide: A V-shaped groove serving to direct a wedge shaped sliding machine element.

Vibrating feeder: A vibratory feeder is an instrument that uses vibration to "feed" material to a process or machine. Vibratory feeders use both vibration and gravity to move material.

Vibrating grizzlies: Bar grizzlies mounted on eccentrics so that the entire assembly is given a forward and backward movement at a speed of some 100 strokes a minute.

Vibrating pebble mill: A size-reduction device in which feed is ground by the action of vibrating, moving pebbles.

Vibrating screen: A sizing screen which is vibrated by solenoid Vibrating Screens are used for screening of rock, ore, coal and similar bulk materials as well as in recycling and waste disposal.

Vibrating screen classifier: A classifier whose screening surface is hung by rods and springs, and moves by means of electric vibrators.

Vibration: A continuing periodic change in a displacement with respect to a fixed reference.

Vibration damping: The processes and techniques used for changing the mechanical vibration energy of solids into heat energy.

Vibration drilling: Drilling in which a frequency of vibration in the range of 100 to 20,000 hertz is used to fracture rock.

Vibration machine: A tool for subjecting a system to controlled and reproducible mechanical vibration. Also known as shake table.

Vibration separation: Separation of grains of solids in which separation through a screen is expedited by vibration or oscillatory movement of the screening mediums.

Vibration suppression: The prevention of undesirable vibration, either through passive means such as damping or through active techniques involving feedback control.

Vibratory centrifuge: A high-speed rotating tool to remove moisture from pulverized coal or other solids

Vibratory equipment: Reciprocating or oscillating devices which move, shake, dump, compact, settle, tamp, pack, screen, or feed solids or slurries in process.

Vibratory hammer: A type of pile hammer which employs electrically activated eccentric cams to vibrate piles into place.

Vibro energy separator: A screen type tool for classification or separation of grains of solids by a mixture of gyratory motion and auxiliary vibration caused by balls bouncing against the lower surface of the screen cloth.

Virial coefficients: For a given temperature T, one of the coefficients in the expansion of P/RT in inverse powers of the molar volume, where P is the pressure and R is the gas constant.

Virmel engine: A cat-and-mouse engine that utilize vane like pistons whose motion is controlled by a gear-and-crank system

Virtual displacement: Any change in the positions of the element forming a mechanical system.

Virtual entropy: The entropy of a system, excluding due to nuclear spin. Also known as practical entropy.

Virtual work: Virtual work is the total work done by the applied forces and the inertial forces of a mechanical system

as it moves through a set of virtual displacements.

Viscoelasticity: Property of a material which is viscous but which also exhibits certain elastic properties such as the ability to store energy of deformation.

Viscoelastic theory: The theory which attempts to identify the relationship between stress and strain in a material displaying viscoelasticity.

Viscous damping: A method of converting mechanical vibrational energy of a body into heat energy.

Viscous fillers : A packaging machine that fills viscous product into cartons.

Volatility: Tendency of a substance to evaporate at normal temperatures.

Volatilization: The conversion of a chemical substance from a liquid or solid state to a gaseous or vapor state by the application of heat, by reducing pressure, or by a combination of these processes.

Volumetric efficiency: Volumetric efficiency (VE) of an internal combustion engine is defined as the ratio of the mass density of the air-fuel mixture drawn into the cylinder at atmospheric pressure (during the intake stroke) to the mass density of the same volume of air in the intake manifold.

Volumetric strain: One measure of deformation. the change of volume per unit volume.

Volute pump: A centrifugal pump housed in a spiral casing.

Von Mises yield criterion: The hypothesis that plastic deformation of a material begins when the sum of the squares of the principal components of the deviatoric stress reaches a certain critical value.

V-type engine: An engine in which the cylinders are placed in two rows set at an angle to each other, with the crankshaft running through the point of a V.

Wagon drill: A vertically accumulated, pneumatic, percussive type rock drill supported on a three- or four-wheeled wagon.

Walking beam: A lever that oscillates on a pivot and convey power in a manner producing a reciprocating or reversible motion; used in rock drilling and oil well pumping.

Walking dragline: A large capacity dragline built with moving feet; disks 20 feet (6 meters) in diameter support the excavator while functioning.

Walking machine: A machine designed to carry its operator over various types of terrain; the operator sits on a platform carried on four mechanical legs, and movements of his arms control the front legs of the machine while movements of his legs control the rear legs of the machine.

Walley engine: A multi rotor engine employing four approximately elliptical rotors that turn in the same clockwise sense, leading to excessively high rubbing velocities.

Wall superheat: The difference between the temperature of a surface and the saturation temperature (boiling point at the ambient pressure) of an adjacent liquid that is heated by the surface.

Walter engine: A multi rotor rotary engine that uses two different-sized elliptical rotors.

Wankel engine: An eccentric-rotor type internal combustion engine with only two primary moving parts, the rotor and the eccentric shaft; the rotor moves in one direction around the trochoidal chamber containing peripheral intake and exhaust ports.

Warm-air heating: Heating by circulating warm air; arrangement contains a direct-fired furnace surrounded by a bonnet through which air circulates to be heated.

Warpage: The action, process, or result of twisting or turning out of shape

Water column: A tubular column situated at the steam and water space of a boiler to which protective devices such as gage cocks, water gage, and level alarms are attached.

Water-cooled condenser: A steam condenser which is for the preservation of vacuum, and in which water is the heat-receiving fluid.

Water heater: A tank for heating and storing hot water for domestic use.

Water tube boiler: A steam boiler in which water flow with in tubes and heat is applied from outside the tubes to produce steam.

Water wall: The side of a boiler furnace consisting of water-carrying tubes which absorb radiant heat and there by prevent excessively high furnace temperatures.

Water wheel: A vertical wheel on a horizontal shaft that is made to revolve by the action or weight of water on or in containers attached to the rim.

Watt's law: Watt's law defines the relationship between power, voltage, and current and states that the power in a circuit is a product of the voltage and the current. Watt 's law has many practical applications, and the formula for measuring Watt 's Law **is P = IV.**

Wave gait: Mode of motion of a mobile robot with several legs in which its components have a wavy motion.

Wave motor: A motor that depends on the lifting power of sea waves to develop its usable energy.

Wedge core lifter: A core-gripping device consisting of a series of three or more serrated-face, tapered wedges contained in slotted and tapered recesses cut into the inner surface of a lifter case or sleeve; the case is threaded to the inner tube of a core barrel, and as the core enters the inner tube, it lifts the wedges up along the case taper; when the barrel is raised, the wedges are pulled tight, gripping the core.

Weight: The gravitational force with which the earth attracts a body

Weightlessness: A condition in which no acceleration, whether of gravity or other force, can be detected by an observer within the system in question.

Well drill: A drill, usually a churn drill, used to drill water wells.

Well-type manometer: It is a glass tube manometer with double leg. One leg has a relatively small diameter, and the second leg is a reservoir; the level of the liquid in the reservoir does not change appreciably with change of pressure; a mercury barometer is a common example.

Wet cooling tower: An

arrangement in which water is cooled by atomization into a stream of air; heat is lost through evaporation. Also known as evaporative cooling tower.

Wet drill: A percussive drill with a water feed either through the machine or by means of a water swivel, to suppress the dust produced when drilling.

Wet engine: An engine with its oil, liquid coolant and trapped fuel inside.

Wet grinding: The milling of materials in water or other liquid.

Wet milling: It is a process in which feed material is steeped in water, with or without sulfur dioxide.

Wet sleeve: A cylinder liner which is exposed to the coolant over 70% or more of its surface. Sleeves with cooling passages can also be known as water-jacket sleeves.

Wet well: A chamber which bring together liquid and to which the suction pipe of a pump is attached.

Wheeled crane: It is a self-propelled crane that rides on a rubber tired chassis with power for transportation provided by the same engine that is used for hoisting.

Willans line: Willan's line method is only used in the compression ignition (C.I) engine. It is not applicable to S.I engine. The line on a graph showing steam consumption versus power output for a steam engine or turbine, frequently extended to show total fuel consumed for gas turbines, internal combustion engines.

Winch: A machine with a drum on which to coil a rope, cable, or chain for hauling, pulling, or hoisting.

Wind age: The deflection of a bullet or other projectile because of wind.

Wind deflection: Deflection caused by the influence of wind on the course of a projectile in flight.

Wind mill: A windmill is a structure that converts wind power into rotational energy by means of vanes called sails or blades.

Wind milling: The rotation of a propeller from the force of the air when the engine is not operating.

Wind power: The removal of kinetic energy from the wind and conversion of it into a useful type of energy: thermal, mechanical, or electrical.

Wind pressure: The whole

force exerted upon a structure by wind. Also known as velocity pressure.

Windup: The twisting of a shaft under a torsional load, usually resulting in vibration and other undesirable effects as the shaft relaxes.

Wire saw: A machine employing one- or three-strand wire cable, up to 16,000 feet (4900 meters) long, running over a pulley as a belt; used in quarries to cut rock by abrasion.

Wobbe index: The Wobbe index is a measure of the interchange ability of gases when they are used as a fuel. It compares the energy output of different gases during combustion.

Wobble wheel roller: A roller with freely suspended pneumatic tires used in soil stabilization

Work: work is the energy transferred to or from an object via the application of force along a displacement. In its simplest form, it is often represented as the product of force and displacement.

Working envelope: The surface bounding the maximum extent and reach of a robot's wrist, excluding the tool tip. Also known as

working profile.

Working load: The maximum load that any structural member is designed to support.

Working pressure: The allowable operating pressure in a pressurized vessel or conduit.

Working space-volume: The volume enclosed by a robot's working envelope

Work-kinetic energy theorem: The theorem that the change in the kinetic energy of a particle during a displacement is equal to the work done by the resultant force on the particle during this displacement.

Work station: A workplace in a production system at which an individual worker may spend only a portion of a working shift.

Work stress: Any external force that acts on the body of a worker during the recital of a task

Wringing fit: It is used in railway **wheels**. A fit of zero-to-negative allowance.

Wrist: It is not one big joint; it has several small joints. A set of rotary joints to which the end effect or of a robot is attached. Also known as wrist socket.

X engine: An X engine is a piston engine with four banks of cylinders around a common crankshaft, such that the cylinders form an "X" shape when viewed from front-on.

X ray machine: It consists of x-ray tube, power supply, and associated equipment required for producing x-ray photographs.

X-ray monochromator: An instrument in which x-rays are diffracted from a crystal to produce a beam having a narrow range of wave lengths.

Y

Yard: A unit of linear measure, equal to 0.9144 meter, 3 feet, 36 inches. Abbreviated yd.

Yardage: A quantity expressed in yards.

Yaw: The rotational or oscillatory movement of a ship, aircraft, rocket about a vertical axis.

Yaw acceleration: Angular acceleration of an aircraft or missile about its normal or Z axis.

Yaw axis: The vertical axis through an aircraft, rocket, or similar body, about which the body yaws; it may be a body, wind, or stability axis.

Yield: The stress in a material at which plastic deformation occurs.

Yield point: The smallest stress at which strain increases without increase in stress.

Yield rate: The quantity of satisfactory material available after the completion of a given manufacturing process expressed as a percentage of the total amount produced.

Yield strength: The stress at which a material shows a specified deviation from proportionality of stress and strain.

Yield stress: The smallest stress at which extension of the tensile test piece increases without increase in load.

Young Helmholtz laws: Two rules describing the motion of bowed strings, the first states that no overtone with a node at the point of excitation can be present

Young's modulus: The Young's Modulus of such a material is given by the ratio of stress and strain, corresponding to the stress of the material. The relation is given below.

Zero bevel gear: A particular form of bevel gear having bowed teeth with a zero degree spiral angle.

Zero defects: A program for recuperating product quality to the point of perfection, so there will be no breakdown due to defects in construction.

Zeroth law of thermodynamics: It states that if two systems are in thermal equilibrium with a third system, then they are in thermal equilibrium with each other.

Zipper conveyor: The belt with zipper like teeth that mesh to form a closed tube; used to handle fragile materials.

Zone: In heating or air-conditioning arrangement, one or more spaces whose temperature is regulated by a single control.

Zone heat: A central heating arrangement to allow different temperatures to be maintained at the same time in two or more areas of a building

Zoom: To expand or decrease the size of an image in an optical system or electronic display.

www.ingramcontent.com/pod-product-compliance
Lightning Source LLC
Chambersburg PA
CBHW071335210326
41597CB00015B/1459